Praise for *Beyond the War on Invasive Species*

"A gathering body of evidence against the scale of chemical interventions in both agriculture and wild nature is fueling a battle of geopolitical proportions. In the process of asking the questions about how best to restore nature, Orion exposes a deep ethical corruption at the heart of both ecological science and the environmental movement."

– DAVID HOLMGREN, from the Foreword

"*Beyond the War on Invasive Species* is a devastating exposé of the military-industrial invasive species complex and a sorely needed and impeccably researched volume that should become one of many as we recover from self-destructive attempts to eradicate parts of nature instead of acting with an understanding of the whole."

– BEN FALK, author of *The Resilient Farm and Homestead*
and founder of Whole Systems Design

"*Beyond the War on Invasive Species* is part of a new, much more nuanced conversation about 'invasive' species that is taking place in science, agriculture, and land management. It provides an analysis of the new science that looks for ecosystem function as well as harm from newly arrived species, looks at species migration in the context of climate change, and broadens our conversation to look at these organisms in the context of the human ecological footprint. Orion offers land-management guidelines, based in permaculture design process, that help to chart a new way forward in our new land of novel ecosystems."

– ERIC TOENSMEIER, author of *Paradise Lot* and *Perennial Vegetables*

"An interesting and valuable contribution to the ongoing refutation of invasive species ideology. Detailed and wide-ranging, Orion extends and deepens several analyses of invasionism, and offers several interesting new perspectives. She points to holistic systems management as an alternative to the current war on invasives. Land managers and invasionists would do well to give it a careful read."

– D.I. THEODOROPOLOUS, author of *Invasion Biology: Critique of a Pseudoscience*

"*Beyond the War on Invasive Species* creates an essential pathway for deeper care of the Earth. The holistic perspective of invasives is shared through deep experience and thoughtfulness and ultimately leads us to a greater and more aligned role in restoration of our home's ecosystems in these changing times. This book offers a critical role in civilization's evolution and highlights actions that recognize deeper values that benefit our society as a whole."

– KATRINA BLAIR, author of *The Wild Wisdom of Weeds:*
13 Essential Plants for Human Survival

"In her fascinating and highly readable book, *Beyond the War on Invasive Species*, author Tao Orion points out the shortcomings of our current approach toward landscape restoration and invasive species. Rather than seeing these exotic plants and animals simply as invaders that need to be eradicated, she argues, we should recognize the beneficial role they play in the environment and the many essential services they could provide to human beings. 'Embracing rampancy,' as Orion exuberantly puts it, turns the perceived problem of invasive species into practical solutions that also allow us to make peace with both the land and ourselves."

— **Larry Korn**, author of *One-Straw Revolutionary: The Philosophy and Work of Masanobu Fukuoka*

"Some of our most productive and tasty plants in the permaculture landscape are vilified as invasive weeds that need controlling. This is a mindset that also promotes a delineation between conservation and agriculture. My personal response is to cultivate fewer conventional annual vegetables and grow and eat as many of these weeds as is appropriate, creating an extensive, diverse, and resilient forage system in my own backyard. It is time to stop putting land management into boxes and create wildlife habitats and food in stacked systems.

"Tao Orion explains how to take advantage of the vigor of 'invasive' edible and useful exotics and harvest them. This is how to bring ecosystems back into balance. This is adaptive permaculture thinking at the broad-scale level. Chelsea Green has produced yet another pioneering book, demonstrating how permaculture is way ahead of conventional land-management practices."

— **Maddy Harland**, editor of *Permaculture* magazine, cofounder of The Sustainability Centre in the UK, and a Fellow of the Royal Society of Arts

"This book brings much-needed balance to the overheated debate about so-called invasive species. Tao Orion's meticulously researched yet engaging work shows that the true culprits are nearly always human-caused disturbance and development, and that species shifts are a symptom, not a cause, of this habitat destruction. *Beyond the War on Invasive Species* is an important book that offers a path away from unsuccessful restoration efforts — based on poor science and policy — and toward new, ecologically sound programs for building and preserving biodiversity."

— **Toby Hemenway**, author of *Gaia's Garden: A Guide to Home-Scale Permaculture* and *The Permaculture City: Regenerative Design for Urban, Suburban, and Town Resilience*

"Tao Orion has brought together personal experience, careful study, and visionary thinking to turn us toward becoming useful people of place. Her exploration widens the narrow concept of invasion (so often repeated but seldom carefully thought through) and elucidates the trouble of shortsightedness. We are not threatened by aliens, but rather we are turning our backs on the big picture."

— **Tom Ward**, author of *Greenward, Ho! Herbal Home Remedies* and cofounder of Siskiyou Permaculture

BEYOND *the* WAR *on* INVASIVE SPECIES

A Permaculture Approach to Ecosystem Restoration

TAO ORION

FOREWORD BY DAVID HOLMGREN

Chelsea Green Publishing
White River Junction, Vermont

Project Manager: Alexander Bullett
Developmental Editor: Brianne Goodspeed
Copy Editor: Deborah Heimann
Proofreader: Brianne Bardusch
Indexer: Peggy Holloway
Designer: Melissa Jacobson

Printed in the United States of America.
First printing June, 2015.
10 9 8 7 6 5 4 3 2 1   15 16 17 18

Our Commitment to Green Publishing

Chelsea Green sees publishing as a tool for cultural change and ecological stewardship. We strive to align our book manufacturing practices with our editorial mission and to reduce the impact of our business enterprise in the environment. We print our books and catalogs on chlorine-free recycled paper, using vegetable-based inks whenever possible. This book may cost slightly more because it was printed on paper that contains recycled fiber, and we hope you'll agree that it's worth it. Chelsea Green is a member of the Green Press Initiative (www.greenpressinitiative.org), a nonprofit coalition of publishers, manufacturers, and authors working to protect the world's endangered forests and conserve natural resources. *Beyond the War on Invasive Species* was printed on paper supplied by McNaughton & Gunn that contains 100% postconsumer recycled fiber.

Library of Congress Cataloging-in-Publication Data

Orion, Tao, 1980–
Beyond the war on invasive species : a permaculture approach to ecosystem restoration / Tao Orion.
pages cm
Includes bibliographical references and index.
ISBN 978-1-60358-563-7 (pbk.) – ISBN 978-1-60358-564-4 (ebook)
1. Permaculture. 2. Restoration ecology. 3. Introduced organisms – Control. 4. Endemic animals – Conservation. 5. Endemic plants – Conservation. I. Title. II. Title: Permaculture approach to ecosystem restoration.

S494.5.P47O75 2015
631.5'8 – dc23

2015006881

Chelsea Green Publishing
85 North Main Street, Suite 120
White River Junction, VT 05001
(802) 295-6300
www.chelseagreen.com

For Sylvan, and the world we give to you.

"We are called to assist the Earth to heal her wounds and in the process heal our own – indeed, to embrace the whole creation in all its diversity, beauty, and wonder. This will happen if we see the need to revive our sense of belonging to a larger family of life, with which we have shared our evolutionary process."

– *Wangari Maathai*

"My friend, all theory is gray, and the Golden tree of life is green."

– *Goethe*

Contents

Foreword

In the 1970s when Bill Mollison and I began working on the concept of permaculture, we recognized local and global biodiversity as a treasure trove of biological wealth that could be combined to create designed ecologies for sustaining humanity beyond the fossil-fuel era. In Australia, the paucity of cultivated indigenous plants and valued native animals was the context in which we highlighted the potential of native plants and animals as integral to permaculture.

As permaculture mushroomed in the context of the 1970s' energy crises and back-to-the-land self-sufficiency, one of the most surprising critiques of the concept was from a small network of botanists and environmental activists promoting the idea that no plants capable of spreading should be grown, anywhere! (At the same time, these activists accepted that agriculture would continue to feed us with foreign species maintained with industrial inputs.) The serious suggestion that permaculture was potentially one of the greatest threats to global biodiversity because of its focus on using a larger, rather than a smaller, range of species to support humanity, seemed remarkably similar to disputes between various schools of Marxism in the 1960s.

Over the following three decades, this ideology of demonizing spreading species as a threat to biodiversity (on a scale rivaling climate change) took over the mainstream of environmental activism and the biological sciences, especially in the English-speaking world. Scientific research papers rebranding spreading species as "invasives" (or in Australia, "environmental weeds") burgeoned, filling peer-review journals. The correct botanical (and emotionally neutral) term "naturalization" was abandoned because it recognized the validity of the process by which species become native to a new place.

This new science of "invasion ecology" informed the education of a cadre of natural-resource-management professionals, supported by taxpayer funds. These resources mobilized armies of volunteers in a "war on weeds." But labor and even machine-intensive methods of weed control were soon sidelined in favor of herbicides that environmentalists and ecologists accepted as a necessary evil in the vain hope of winning the war against an endless array of newly naturalizing species. For the chemical corporations, this new and rapidly expanding market began to rival the use of herbicides by farmers, with almost unlimited growth potential, so long as the taxpayer remained convinced that the war on weeds constituted looking after the environment. In Australia, the visionary grassroots Landcare movement, started by farmers in the early 1980s, was reduced to being the vehicle for implementing this war on weeds.

The criticism of permaculture by the environmental orthodoxy was not due to the scale of any real impacts of permaculture, but more because of the perceived audacity of using ecological arguments to justify the use of a wider range of species not indigenous to a site or even a bioregion. Permaculture practitioners were mostly doing little more than attempting to maintain the lineage of agricultural and horticultural research into promising species, as governments beguiled by economic rationalism abandoned their responsibilities to invest in economic botanical research. Most permaculture teachers and designers accepted the findings of invasion ecology at face value and sought to minimize risks of unintended naturalization.

For me, this pragmatic accommodation drove permaculture away from the principle of working with rather than against nature. My own interest in abandoned gardens and arboreta and rewilded farms and urban places as sources of permaculture inspiration was intensified through working with planner and resource ecologist Haikai Tane in New Zealand in 1979 and 1984. We coined the term "ecosynthesis" to describe the relatively rapid restoration of ecosystem function that we saw in the recombinant mixtures of native and foreign species that colonized abandoned landscapes. We also recognized how this process was generating new resources that could support human populations beyond the fossil-fuel era. Further, we recognized that these novel ecosystems were the best models for the design of intensively managed human settlements. Beyond this, Tane branded the war against naturalizing species as nativism, an ideology that sought to separate nature into good and bad species according to some fixed historical reference.

As Orion makes clear in this excellent review of the application of invasion ecology to the practice of ecological restoration, much of the dysfunction can be traced back to the triumph of reductionism over a more holistic systems approach in the biological sciences. It is a great irony that ecology, the scientific discipline that was founded on holistic understanding, was overwhelmed by reductionism in the 1980s. This coincided with the flow of cheap oil from the Alaskan North Slope and the North Sea, the dismissal of the inconvenient truth of *The Limits to Growth*, the rise of economic rationalism, and the demonizing of the counterculture, all aspects of the Thatcherite-Reaganite revolution that spread from the Anglo American countries in the 1980s.

My experience in articulating ecological arguments that naturalization of species could maintain and rebuild ecosystem services seemed like reputational suicide among environmentalists and ecologists. Pointing out that our views were in line with the UN biodiversity convention (1992) didn't help (even though it recognizes the validity of conserving biodiversity wherever species are wild). Even the more modest case, that major efforts at removal of established species would do more harm than good, was dismissed.

It was not until the turn of the millennium with the aid of the Internet that I became aware of the growing number of ecologists who were questioning this orthodoxy. In the book *Invasion Biology: Critique of a Pseudoscience* (2003), independent Californian naturalist David Theodoropoulos went further, claiming that the spread of species in a rapidly changing world would do more to conserve biodiversity than the massive efforts to reverse naturalizations and protect species in collapsing niches. Since then, a growing body of peer-reviewed research in the field of what is now called "novel ecosystems research" is providing the evidence toward a tipping point that may reform the field of invasion ecology. I believe abandoning emotionally loaded and unscientific terms such as "invasive" and "weed" will be a symbolic and necessary step in the process. Those driving the restoration industry (primarily in affluent countries) are mostly yet to recognize the shift in the science, or the leakage of disillusioned restoration professionals who have responded to the bizarre contradictions in the practice of restoration ecology by adopting more holistic responses to ecological disturbance and species naturalizations, such as permaculture. In this way, the industry is progressively losing its most motivated and ethical practitioners.

I see the importance of Orion's book as threefold. The book traces the story of how well-intentioned concerns about biodiversity loss and ecological change were captured and corrupted by corporations selling chemical solutions to the perceived problem of "invasive species." Using measured language and open questions, Orion allows the ordinary reader to judge this process. A gathering body of evidence against the scale of chemical interventions in both agriculture and wild nature is fueling a battle of geopolitical proportions. In the process of asking the questions about how best to restore nature, Orion exposes a deep ethical corruption at the heart of both ecological science and the environmental movement.

Perhaps more fundamentally, the author's personal experience as a dedicated and innovative restoration practitioner, speaking directly to both her peers and the general public in a reflective tone, provides a great model that can help all of us move farther along that long, winding road of learning how to work with nature rather than against her.

Finally, as the baby boomers who rode the first wave of modern environmentalism fade from the scene, I find it significant that it is a young woman who provides the hope that environmentalism, having lost its way through a futile and destructive war against nature, can return to the path we collectively lost.

DAVID HOLMGREN

Work Song: A Vision

If we will have the wisdom to survive,
to stand like slow-growing trees
on a ruined place, renewing, enriching it
if we will make our seasons welcome here,
asking not too much of earth or heaven,
then a long time after we are dead
the lives our lives prepare will live
here, their houses strongly placed
upon the valley sides, fields and gardens
rich in the windows. The river will run
clear, as we will never know it,
and over it, birdsong like a canopy.
On the levels of the hills will be
green meadows, stock bells in the noon shade.
On the steeps where greed and ignorance cut down
the old forest, an old forest will stand,
its rich leaf-fall drifting on its roots.
The veins of forgotten springs will have opened.
Families will be singing in the fields.
In their voices they will hear a music
risen out of the ground. They will take
nothing from the ground they will not return,
whatever the grief at parting. Memory,
native to this valley, will spread over it
like a grove, and memory will grow
into legend, legend into song, song
into sacrament. The abundance of this place,
the songs of its people and its birds,
will be health and wisdom and indwelling
light. This is no paradisal dream.
Its hardship is its possibility.

— *Wendell Berry*

Introduction

Several years ago, I found myself on the front lines of an unexpected war. It was 2010, and I'd been hired as a botanist for the Lane County Department of Public Works in Oregon's Willamette Valley, where I live and farm with my family. My job was to identify plants, collect seeds, and strategize innovative ways to manage an emerging wetland restoration site located on county land. The site had been farmed for more than fifty years, most recently for ryegrass seed production, which the Willamette Valley is famous for. The county had purchased the 64-acre site as part of a mitigation plan – mitigating the effects of expanding an adjacent landfill onto intact wetland. The goal of the project was to convert the site back to a wetland – similar to what it might have looked like before agriculture, ranching, logging, and mining altered the ecological character of the region. At the time, I thought this was a valuable endeavor. Returning a small portion of the valley dominated by grass seed farms to wetland habitat seemed like a noble, if inadequate, attempt to integrate more habitat and ecological function into a landscape where it had largely been lost.

Before taking this job, I thought of restoration as an important undertaking, but I didn't understand what "restoration" meant in a modern context. I knew it implied a process of improving habitat and biodiversity, but I was unfamiliar with how organizations developed projects to achieve these goals. I didn't realize that there's a subtext to restoration, as commonly practiced, of native species as "good" and invasive species as "bad" and that this dichotomy is the principle that underpins many of the decisions and strategies driving restoration projects around the country. The Society for Ecological Restoration defines restoration as "the process of assisting the recovery of an ecosystem that has been degraded, damaged, or destroyed." My assumption was that restoration would align with this sentiment not only in theory, but in practice.

Needless to say, my training is not in the field of ecological restoration per se. It's in farming, agroecology, permaculture design, and ethnobotany. Before I started the job, I was familiar with invasive species as a land manager, organic farmer, and permaculture designer, but not as someone trained in restoration ecology – and I didn't realize how different these fields would be. But I knew, for example, how frustrating the spiny vegetation of Canada thistle can be when it encroaches in a bed of peppers, and I have found myself confounded by the persistence of field bindweed. I know how rapidly the long thorny canes of Himalayan blackberry creep across unmanaged pasture, and I have received my share of scratches in attempts to cut them back.

But I never considered invasive species to be threatening or loathsome. I thought of their proliferation mostly as a management issue – the result of either too much, as in the case of Canada thistle spreading on overgrazed pastures, or too little, as with blackberries that grow on field margins and other underutilized spaces – usually in highly modified ecosystems. I credit permaculture and organic farming with teaching me that the proliferation of invasive species represents an opportunity to understand ecological dynamics at a deep level and to manage land based on that understanding.

Specifically, organic farming taught me how to manage these plants using organic methods: mowing, cultivating, burning, and mulching. Permaculture taught me that every organism is intrinsically connected to the ecosystem of which it is part and that addressing the ecological conditions that promote a species' proliferation, whether you wish to discourage or encourage its survival, will enhance the ecosystem overall. I knew that invasive Scotch broom is a hardy, drought-tolerant, nitrogen-fixing plant that grows where soil has been disturbed. Blackberries and Canada thistle also find their homes on land where conditions are less than ideal for native or other desirable vegetation.

Admittedly, I came to restoration with a bias, or at least a sense that every organism in an ecosystem is an indication of complex underlying ecological dynamics, whether it is considered native to an area or not. So when I learned that the use of herbicide is common in the field of ecological restoration as a way to manage invasive species, I was – quite frankly – appalled. When I accepted the position with the Lane County Department of Public Works, the restoration project was already two years into a five-year plan. The first three years involved annual broadcast spraying of Roundup – known as "nuking" in restoration lingo – across the site. According to the management plan, this regular herbicide application creates a "moonscape" – another common term in the restoration lexicon – that supposedly makes it easier for native plants to survive when they're later sown and planted. After this "nuking," invasive species are spot-sprayed as they crop up.

From California to South Africa, New York to New Zealand, invasive species seem to be everywhere, their populations expanding and threatening ecological integrity around the world. A 1998 Princeton University study found that invasive species are the second greatest threat to global biological diversity.[1] In 2011, the International Union for Conservation of Nature, the world's oldest and largest global environmental organization, warned that invasive species wreak havoc on economies and ecosystems.[2] Organizations such as the National Wildlife Federation and the National Audubon Society consider invasive species to be serious impediments to healthy wildlife habitat and the survival of endangered species.[3] And government agencies, including the US Forest Service and Bureau of Land Management, blame invasive species for losses and permanent damage to the health of natural plant communities.[4]

Species invasions are also costly. According to The Nature Conservancy, worldwide spending on invasive species totals $1.4 trillion every year, equal to 5 percent of the global economy.[5] The United States alone spends $137 billion annually to contend with them.[6] The National Invasive Species Council holds invasive species accountable for "unemployment, damaged goods and equipment, power failures, environmental degradation, increased rates and severity of natural disasters, disease epidemics, and even lost lives."[7]

Given the apparent threats posed by invasive species, it makes sense that their eradication has become a central organizing principle of the practice of restoration. After all, if they are perceived as degrading ecosystems, then the practice of restoration as assisting in the repair of degraded ecosystems should focus on their elimination. I know that people who work in restoration care deeply about the loss of habitats, loss of ecological function, and declining biodiversity that are readily apparent in seemingly every ecosystem on Earth. Invasive species in many cases are part of this trend, and while I agree that invasive species are less ideal than the diverse and robust native flora and fauna they appear to dominate and replace, invasive species themselves aren't the actual problem; they are merely a symptom. I remove invasive species on my own property and in my own design work, but always in the context of creating something greater, more robust, diverse, and resilient than what was there before, and I have never considered using herbicides to manage them. As individuals, we have to take responsibility for the land. We have to draw the line.

Herbicides are favored as a restoration tool because they represent a relatively cheap, readily available, and supposedly benign method of killing certain plants and favoring others. Herbicides can be applied in small, targeted amounts, ridding the site of unwanted species with relatively little disturbance (at least visually) compared to seemingly more destructive eradication tools like bulldozers and tractor-mounted mowers. Planes, helicopters, and people with backpack-mounted spray tanks can apply herbicide in remote locations that are not easily accessed by other eradication equipment. Herbicide application can be selective, as it is easily applied to certain plants and not others on a small scale. Working with a weed whacker or machete does not allow for the same careful treatment. In some cases, such as the Lane County project I worked on and others I visited, herbicides are used on a larger scale in the hopes that the site can start with a "clean slate," leaving the desired native vegetation more or less free from the competition expected from rampantly growing invasive species.

I understand the need for cheap and straightforward vegetation management strategies, especially on a large scale. I've spent many a sweaty hour at the helm of a well-sharpened hoe. But the idea that herbicides are considered a necessary part of restoring an ecosystem didn't sit right with me. If organic farmers can grow nonnative plants like broccoli, potatoes, and turnips without the use of chemical pesticides, why should native plants need them, I wondered? I assumed that native plants and animals are supposedly more suited to local conditions than nonnative ones, so it seemed to me that they should be easier to establish and maintain than those crops commonly grown in a farm setting.

As I considered how to curb the rampant growth of invasive species without chemical intervention on the site I worked, I turned to organic farming for possible solutions. At the first organic farm I worked on, we'd used backpack-mounted propane torches for spot-burning the carrot beds. Carrots take a couple of weeks to germinate, and many weeds emerge sooner, impeding their growth. With a few well-timed passes of the torch, I could drastically reduce the number of weeds competing with the carrots, making successive weeding more efficient.

I reasoned that a similar approach could be taken at this wetland restoration site – eliminate the early germinating and rapidly spreading invasive plants, and make room for more slowly emerging desirable native species. Although backpack burning is standard in organic farming, it had not been used as a restoration tool in the Willamette Valley. Still, my supervisors were willing to let me try it.

We found that spot burning worked successfully to eliminate populations of rattail fescue and worked reasonably well on false dandelion and the pennyroyal that grew along the pond margins. We sowed seeds of native tufted hairgrass and American sloughgrass into the burned patches, which germinated readily in the moist soil. This approach seemed to work well, at least as well as spot spraying.

However, as I stood out in the middle of the site one cool, windless morning in April, eight months into the project, with my propane torch in hand, I burned a native spotted frog – an amphibian recently listed as a threatened species in Oregon – that had been hiding in a patch of the invasive rattail fescue. Its skin was badly charred, and it died soon after. I couldn't help but feel that my actions were not serving to enhance the ecosystem in the best way possible. I was glad to keep herbicides out of the watershed, but there had to be better ways to improve ecological functionality and habitat value than by removing certain plants, especially if their eradication caused harm to the very creatures that were supposed to benefit from restoration of the land. This was the same problem I had with herbicides: No matter what, the ends justify the means. Spot burning invasive plants was supposed to be a tool that would create more and better habitat for native frogs and plenty of other creatures over time. Did burning a few frogs along the way matter in the context of the long-term goal of creating a wetland ecosystem on the site?

These questions haunted me as I continued to work on the project. I began stamping around in the patches of rattail fescue and pennyroyal, hoping to scare the frogs out of their habitat before I ignited the torch. But then I realized that the patches were everywhere and that the pennyroyal, rattail fescue, and false dandelion were serving other valuable roles in the ecosystem of the project site. The false dandelion and pennyroyal were covered with honeybees and native pollinators when they bloomed, and these too fell as ashes under my torch, leaving the pollinators to forage elsewhere.

I realized that even though I was using a technique borrowed from organic farming instead of herbicide-based eradication, I was still working within a paradigm that viewed invasive species as threats to ecological health when my assessment of reality on the ground demonstrated the opposite. The invasive species were serving ecological functions – the bees didn't appear to mind that the nectar they sipped came from a flower that originated in Europe nor did the frogs seem to care that the low-growing thatch of rattail fescue hailed from the same region. They may have preferred native vegetation, but it was hard to tell, especially since native plants were not thriving and certain invasive species were. Was the success of the invasive species based on the fact that they were supposedly more aggressive or more competitive, or were other factors contributing to their proliferation? And were the pollinators and frogs just making do with what was available, or were their pollination activities reflective of a changed and changing ecosystem? I realized that in order to develop a truly holistic approach to restoration, one in which the ends and the means were justified by enhancing ecological integrity every step of the way, I had to take a step back and look at the bigger picture, taking what ecologist H. T. Odum called a "macroscopic" view of observed ecological phenomena.

So I decided to zoom out from the frogs, the grasses, and the ponds and look at the site in terms of its own history, as well as how it was situated within the changing dynamics of the surrounding ecosystem. The 64-acre site had been farmed for more than fifty years and probably had been a wetland within its long history, prior to the construction of flood control dams and the massive wetland draining efforts throughout the late nineteenth and early twentieth centuries that enabled the development of tillage-based agriculture in the Willamette Valley. From wetland to farm to wetland again, each of these transformations engendered ecological changes, some more obvious than others, in soil characteristics, species assemblages, water flows, and other dynamics.

The plan to restore the site to wetland characteristics had called for the removal of 30,000 dump truck loads of soil – enough to fill a football stadium 330 feet deep – in order to reach the water table and create year-round standing water. Bulldozers, scrapers, excavators, and dump trucks rolled over the site day after day, for months on end, digging, shaping, and compacting the soil. The site was sprayed and sprayed again. Thousands of pounds of native seeds were broadcast across the site, and yet invasive species still thrived.

In fact, the treatment of the site during the course of its restoration, as well as its recent history as a grass seed farm, created the niches for invasive species to thrive. The soil excavated from the site was mostly organic matter–rich topsoil that was trucked away to cap the adjacent, nearly full landfill. Wide swathes of false dandelion radiated out from the imprints of the tire tracks in the bare clay subsoil, and pennyroyal spread along the wet margins of the newly excavated ponds where nothing else grew. False dandelion, with a taproot like its more famous cousin, will over time break up the most compacted soils. Pennyroyal, though it forms dense mats, attracts pollinators, which in turn feed birds, amphibians, and fish, each in their way contributing to the development and diversification of the emerging ecosystem.

And though thousands of dollars' worth of native seeds were purchased and sown on the site, relatively little germinated. Native wetland plants are not adapted to bulldozer tracks and compacted clay subsoil. Restoration in this case entailed fabricating a novel ecosystem, so from an objective standpoint, the proliferation of novel species could be considered an expected outcome. However, the restoration plan was based on the concept that once the physical environment (in the form of a wetland) was created, and the pesky "weeds" removed, native species would flourish on the site. Were the native plants we chose adapted to early-, mid-, or late-succession wetlands? As far as I could tell, we were mixing plants from all different parts of the ecosystem succession spectrum, hoping that some of them would take root. I suggested that we choose seeds from native plants that were similar in form and habit to the invasive species, reasoning that the invasive species were filling a functional role in the ecosystem on some level and that if we wanted native species to flourish instead, we should mimic the characteristics of the invaders.

We decided to give it a try. We chose native self-heal and mat-forming checker mallow to mimic the pennyroyal. We sowed tufted hairgrass, American sloughgrass, and various sedges to fill the void of the rattail fescue. We planted yampah tubers and bulbs of native lilies including camas, narrowleaf onion, and harvest lily to mimic the decompacting action of the false dandelion's deep taproots. In the spring, the wetland bloomed with flowers, and pale green grasses and sedges sparkled on the pond margins.

In order to stock the site with the native plant species we were looking for, we scouted the region for robust populations of native plant seeds, bulbs, and tubers. We harvested 30,000 camas bulbs from a location where a new road was being constructed. We collected wapato tubers from a local farmer who happily plowed up the mucky ground where they grew in the hopes of expanding his grass seed operation. We picked seeds from the endangered Bradshaw's lomatium growing at the landfill and Oregon gumweed, tule, and water plantain from roadside ditches. Mule's ears and tarweed grew next to the interstate, their full seed heads whipped back and forth by passing traffic.

It was difficult to reconcile the irony that we found the largest stands of native seeds, tubers, and bulbs – which were so critical to "restore" to the wetland site – from places where they were obviously unwanted or unutilized. As I thought more deeply about this conundrum, I realized that it was the heart of the issue. The native plants, which have been, and continue to be, cultivated and valued as food, medicine, and fiber crops by the indigenous people of our area, are marginalized by most everyone else. If restoration of native species and habitats is critical enough to warrant the use of herbicides to maintain and enhance their populations, then why doesn't our culture value these organisms and the ecosystems where they thrive? Native plants and animals should be valued in agricultural, forestry, mining, and ranching operations; shopping malls, golf courses, and power plants should be habitats and havens. But, for the most part, this isn't the case. Restoration focuses on native species in certain isolated contexts – like the 64-acre wetland where I worked – but not in others. It typically occurs in so-called natural areas that are artificially separated from the majority of "productive" ecosystems where less than optimum ecological outcomes are considered normal.

It occurred to me that if we are to restore and enhance populations of native species, then we must restore our sense of belonging within the ecosystems that we depend upon. We must reimagine restoration as a practice that takes place in all ecosystems, especially those from which we derive our daily needs. Ecosystems should be restored and enriched by every action we take and decision we make, including the methods we use to procure our food, shelter, water, and other necessities of daily life. Restoration should be designed into every facet of our lives.

It is here that permaculture design, rooted in the close study of dynamic living systems, offers a unique set of solutions to a series of complex and seemingly intractable challenges. Although permaculture design is usually associated with farming and gardening, its ethics and principles can and should be applied in any context – from urban planning to alternative currencies to democratic decision making – and best practices for enhancing ecosystem structure, function, and diversity are especially well-served by these ideas. The movement's motto – *Ingenio patet campus* – means "the field lies open to the intellect." Permaculture design theory offers a framework for thinking of restoration not just in terms of eliminating invasive and favoring native species, but by providing a roadmap of how to integrate ecologically restorative practices into home landscapes, cities, suburbs, farms, and forests.

Based on close observation of natural systems, permaculture design theory teaches that everything gardens – that everything from a bulldozer to a bullfinch to a bull thistle is affected by and affects its surroundings. From this whole systems–based perspective, the proliferation of a particular invasive species is not simply a quality of the organism itself, but a reflection of the ecosystem where it is found. Because permaculture design is focused on the relationships among elements rather than on considering an element in isolation, it offers a unique way to understand how rampancy of a particular species might be indicative of larger processes at work. It is interdisciplinary in the broadest sense of the word, examining the relationships between ecology and economics, politics and history, design decisions and land use planning, as well as the relationships between people and the ecosystems that sustain them. And it moves from patterns to details, making assessments on how big picture patterns such as climate, geology, site history, and offsite influences like flows of wind, water, nutrients, and pollution affect the details of an ecosystem, whether it's shifting populations of plants and animals or the ways that water moves and soil profiles change over time. Because it's rooted in this kind of thinking and decision making, permaculture design offers a unique framework to understand invasive species and engage in a more holistic, and more successful, restoration planning process.

Permaculture design also works in real time and plans for the future. Restoration as it is commonly practiced is based on the concept that an ecosystem can or should be brought back to a previous time or historic species assemblage, and it follows that invasive species would be considered impediments to these ends. But turning back the clock on the ecosystem evolution is not only impossible; it distracts us from understanding how the ecosystems of today are responding to real and serious challenges. Permaculture design gives us a path forward, to the future.

The process of permaculture design is ethically based, meaning that every action and decision should evaluate how best to care for the earth, care for people, and reinvest surplus into regenerative systems. This offers an important advantage over more conventional approaches to managing invasive species, which can include the use of herbicides and other ecologically destructive practices. It factors in environmental harm, health effects on people, ecological or social benefits, and financial support of large agrichemical corporations as part of the decision making around best land management practices.

Permaculture design ethics and principles offer simple but elegant ways of conceptualizing the relationships of dynamic living systems from understanding the time-based nature of how they change to the multiple functional relationships present between the living and nonliving elements they contain. In this light, the idea of "invasive species" is peculiar since all plants and animals are native to our singular and unique planet. Bill Mollison, co-originator of the permaculture concept, states, "I use only native plants, native to the planet Earth. I am using indigenous plants; they are indigenous to this part of the Universe."[8]

This doesn't mean that permaculture-based restoration entails a laissez-faire approach to land management. Instead, the solution-oriented practice of permaculture promotes the creative utilization of invasive species as a by-product of holistic landscape planning. Invasive species have stories to tell and information to offer, and permaculture design strategies offer tools to interpret them in the context of moving ecosystems toward more desirable ends. Instead of inciting fear or frustration, perhaps we can come to see the presence of invasive species as an exciting opportunity to engage with the land in deep and meaningful ways.

And active engagement is exactly what we need in order to curb the loss of ecological integrity and global biodiversity as the climate changes and other ecological pressures intensify. The ecosystems that we live in and depend upon require our thoughtful, diligent, and protracted participation within them to ensure their resilience. Restoration as it is often practiced aims for this by attempting to stem the tide of invading species, but it too often misses the mark by failing to address the wider ecological context of which invasive species are part. We must widen the focus of restoration to encompass the human-designed and human-driven processes that are leading to declines in ecological function and biodiversity. When we do so, we will bring the restoration process home, to farms, forests, cities, and suburbs. The design of each of these has ecological ramifications that ripple through ecosystems near and far. In their redesign lies the solution to the challenges they present.

CHAPTER 1

Against All Ethics

"Man's attitude toward nature is today critically important simply because we have now acquired a fateful power to alter and destroy nature. But man is a part of nature, and his war against nature is inevitably a war against himself. [We are] challenged as mankind has never been challenged before to prove our maturity and our mastery, not of nature, but of ourselves."

— RACHEL CARSON

Invasive species are nothing new. Farmers have dealt with weedy plants for as long as they have tilled the soil. Darwin noticed that some species appeared more competitive than others, and later, he used their presence to strengthen his theory of the influence of natural selection on evolution. But although the proliferation of novel species has been observed for centuries, widespread concern has only emerged within the last several decades. In 1958, Charles Elton, the father of invasion ecology, published *The Ecology of Invasions by Animals and Plants*, which argued that organisms flourishing in regions where they did not evolve should be considered invaders that pose imminent harm to their introduced ecosystems. He wrote that there were "two rather different kinds of outbreaks in populations: those that occur because a foreign species successfully invades another country, and those that happen in native or long-established populations," positing for the first time that organisms relocating from "foreign" lands act differently from those that moved around in their "natural" ecosystems.[1]

Elton, an animal ecologist, lived through World War II in England, where he worked to ensure that England's strictly rationed food supplies were safe from rodents, including three invasive species of rats and mice that had been introduced to Europe in the eleventh century. Prior to the publication of *The Ecology of Invasions by Animals and Plants*, Elton wrote extensively about succession, dispersal, and other processes of ecosystem transformation.[2] His postwar publications departed from this early work and focused instead on the apparent dangers of so-called invading species. He wrote: "It's not just nuclear bombs and war that threaten us. There are other sorts of explosions, and this book is about ecological explosions."[3]

Elton wrote these words on the cusp of a major transformation in global ecosystems. After World War II, the US government undertook a strategic mission to secure food supplies in war-torn and famine-prone Europe and Japan, as well as to provide technology to stabilize food supplies throughout the world as a means of quelling further social upheavals.[4] The discovery and adoption of synthetic nitrogen fertilizer earlier in the century enabled farmers to cultivate more land without having to grow the cover crops or provide pasture for animals that once provided fertility. Technologies used in the production of tanks and airplanes found new uses in agriculture as tractors and pesticide sprayers. As the size and scope of food production grew, so too did farmers' reliance on pesticides, some of which were repurposed and reimagined from their wartime applications like nerve gas into potent organophosphate insecticides like parathion and malathion. Developments in the synthesis and manipulation of organic molecules also prompted the discovery of organochlorine molecules including DDT, aldrin, and dieldrin.

Another legacy of World War II was the development of herbicides. During the war, scientists in England, Germany, and the United States researched chemicals that could destroy crops and food supplies of enemy forces. Although they were never deployed during World War II, researchers studied molecules including 2,4-D and 2,4,5-T for their potential applications as chemical weapons. The British military first tested the wartime application of herbicides in Malaya (the Philippines) in the early 1950s to eliminate food crops of communist revolutionaries. The US military used a similar approach on a much larger scale when it applied nineteen million gallons of herbicides, including eleven million gallons of Agent Orange (a mixture of 2,4-D and 2,4,5-T primarily manufactured by Monsanto and Dow Chemical) in Vietnam beginning in 1962.[5]

Herbicide use in agriculture and other industries increased dramatically during the mid-twentieth century. The rapid pace of industrialization and housing development in the postwar United States required large supplies of timber. Forest clearing in the United States and globally intensified to meet demand; over time, herbicide use became standard practice among foresters who sought to maximize timber yields and eliminate competing vegetation. Public scrutiny of 2,4,5-T (which was banned by the Environmental Protection Agency in 1985) and 2,4-D increased following the Vietnam War, and several other herbicides marketed as safe alternatives emerged on the market, including glyphosate and triclopyr.

As agriculture expanded and deforestation intensified with the availability of Green Revolution technology, postwar travel and trade also increased dramatically. Commercial airline flights became popular, making intercontinental travel possible on a much larger scale than at any time in history. Plants and animals from exotic places proliferated in a thriving nursery trade. Elton's observation of an ecological explosion was a prescient assessment of a global ecological transformation due to development and the intensification of agriculture in the postwar years. Though Elton was the first to highlight the ecological role of invasive species in changing ecosystems, he wasn't the first or the only one to notice that many species seemed to be out of place.

Invasions and Industry Interests

Invasive species appeared in the wake of industrial development in the United States, and some of them began to have a negative impact on agriculture, forestry, ranching, and manufacturing. For example, novel weeds spreading across ranch lands reduced available forage and reduced cattle weights, and birds like starlings ate from grain stores destined to fatten cattle. In 1900, the US Congress passed the Lacey Act, which made it unlawful for any person to import undesirable species such as the starling, fruit bat, mongoose, and other species that would be "injurious" to the interests of agriculture or horticulture. In 1974, Congress broadened the scope of this legislation into the Federal Noxious Weed Act, which bans the importation of plants that interfere with the growth of "useful plants," that is, those that are deemed detrimental to agriculture or commerce. When novel organisms started affecting business interests, they started to be characterized as "noxious."

However, some of today's "worst" invasive species were intentionally introduced for ecological benefit. Prior to being classified as invasive, state and federal agencies recommended plants like sea buckthorn, bush honeysuckle, and multiflora rose for their usefulness as habitat or food for wildlife. From 1935 through the 1950s, the Soil Conservation Service distributed eighty-five million kudzu seedlings to stabilize eroding slopes in the Southeast and paid landowners $19.75 per hectare to plant the rapidly growing vine.[6] Scotch broom was sold as an ornamental and planted along roadsides and cut banks throughout the western United States during the 1940s and 1950s. Scotch broom, kudzu, and other novel species proliferated in their new ecosystems, occasionally spreading rapidly and densely enough to warrant concern that they were displacing native species or interfering with other interests.

Some "invasive" animals share a similar history of unintended consequences following their deliberate introduction. Sugar cane plantation owners introduced mongooses to Hawaii starting on the Big Island in 1883. Although they were introduced in the attempt to control rats feeding on sugar cane, the mongoose preferred feeding upon the nene, a native Hawaiian flightless goose. Cane toads were introduced to Australia from North America in the 1930s to feed on cane beetles, although it was soon discovered that the toads could not climb to the top of sugarcane plants where the beetles lived. Mute swans were introduced from Europe as ornamental species for northeastern park ponds in the late nineteenth century before people realized how territorial they are when they nest, driving other birds from the areas where they guard their eggs.

Other introductions were unplanned, such as in the early 1890s when spotted knapweed seeds hitchhiked in bales of European alfalfa and became established in western rangelands. In 1985, zebra mussels were brought to the United States in the ballast water of oceangoing freighters from Europe and now make their home in the Great Lakes region. And some invasive species like purple loosestrife have escaped from gardens and plant nurseries where they were intended as ornamentals and now thrive in wetlands and other ecosystems throughout the country. Of the approximately 4,500 intentionally and unintentionally introduced species with self-sustaining populations – those that can survive outside of intentional cultivation – in the United States, 122 are officially recognized as "harmful" according to federal law.[7]

Fanning the Flames

Over time, concern over these species has increased, and the focus of managing invasive species has moved from import limitations, fines, and penalties to herbicide-based eradication control measures – a change in focus likely influenced by business interests. In 1999, President George W. Bush signed Executive Order 13112, creating the National Invasive Species Council (NISC) and the Invasive Species Advisory Committee (ISAC), a consortium of academic, professional, and industry representatives that make invasive species management policy recommendations to the national council. Executive Order 13112 mandates the federal government to "provide for restoration of native species and habitat conditions in ecosystems that have been invaded," establishing a legal precedent for the relationship between species invasions and restoration. The federal government allocates upwards of $1 billion annually to research and control invasive species, nearly half of which is spent on "control" measures, including pesticide applications.[8]

However, as late as 2005, the NISC had yet to agree on a clear definition of what constitutes an invasive species.[9] In 2006, they published an eleven-page document that explains the ins and outs of what is meant by the phrase *invasive species*. Since the term is subjective both in its origin and application, the council offers several examples of how and why a species fits the definition. According to the NISC, invasive species can decrease property values, alter business opportunities, compete with people for water, cause declines in timber availability, change wildfire frequency (which harms property or affects range or timber value), decrease carrying capacity for livestock on rangelands, and cause declines and alterations in sport fishing and recreational opportunities.[10] Although the council acknowledges the list is not comprehensive, the definition of what constitutes an invasive species is a subjective reflection of human interests.

The NISC is modeled after an organization launched in California in 1992, first known as the California Exotic Pest Plant Council, and later as the California Invasive Plant Council (Cal-IPC). This group, which has served as a model for similar groups in other states, brought together numerous stakeholders from government agencies, nonprofit conservation organizations, and – alarmingly – manufacturers of herbicides and spray equipment. Monsanto has sponsored Cal-IPC since its inception, and both Dupont and Dow AgroSciences have also supported the group.[11] In 1993, organizers of the first Cal-IPC conference, which was focused on the management of invasive giant reed, welcomed Dr. Nelroy Jackson, a

Monsanto employee who helped develop the glyphosate-based Roundup and Rodeo herbicides specifically for the "weed management in roadside, brush, aquatic and habitat restoration markets."[12] Jackson's paper stated that "chemical weed control is the optimal method for control and removal of exotic plant species during the establishment and improvement of most native habitat restoration projects."[13] He suggested the use of both Roundup and Rodeo to control giant reed.

As an employee of Monsanto, Dr. Jackson served as a technical advisor to Team Arundo – a task force in charge of strategizing ways to deal with giant reed in southern California waterways. He served as president of the California Weed Science Society, director of the Western Society of Weed Science, and on the board of the Weed Science Society of America. He also helped to develop the executive order signed by President George W. Bush and the resultant management plan for the federal Invasive Species Advisory Committee, which he later served as vice-chair.[14]

Jackson's represents one of many close and troubling relationships between the giants of the pesticide industry and policy development on invasive species management. Joe Yenish, an employee of Dow AgroSciences is the president-elect of the board of directors for the Western Society of Weed Science for the 2014–2015 term. The Weed Science Society of America, which promotes "research, education, and awareness of weeds in managed and natural ecosystems," features a representative from Syngenta and one from Dow Chemical on their board of directors. Their "presidential sustaining members" – those that donate at the highest level to the organization – include Monsanto, Syngenta, Dow Chemical, Bayer CropScience, BASF Corporation, and DuPont.

In 2011, the annual conference of the Weed Science Society of America included presentations such as "Control of Leafy Spurge with Imazipic and Saflufenacil Applied in Spring," "Aminocyclopyrachlor Herbicide Mixtures for the Western US Vegetation Management Market," "Tribenuron Enhanced Control of Yellow Foxtail (*Setaria pumila*) with Flucarbazone," and "Native Grass Establishment with Aminopyralid."[15] Over half of the nearly two hundred presentations at this symposium investigated plant responses to herbicides, while only four investigated organic methods. Thirty-five of the studies were conducted in collaboration with industry representatives from Dow Chemical, DuPont, Syngenta, and Monsanto.

Corporate interests are also represented in government-level restoration planning. Hugh Grant, the CEO of Monsanto, sits on the board of Tetra Tech, which consults on writing environmental impact reports for the Bureau of Land Management, a publicly funded federal agency. In Oregon, the Bureau of Land Management currently approves the use of four different herbicides and is "examining the potential for using" fourteen more for invasive species management on its extensive landholdings.[16]

It gets worse. In the nineteenth century, the United States granted 17.5 million acres of federally owned land to states for the establishment of publicly funded land grant universities. This land and the universities now sited on it form the backbone of agricultural research in the United States and provide benefits to students and farmers alike from their extension offices, publications, and educational offerings. Every state has at least one land grant college or university. These universities are also home to agricultural experiment stations and cooperative extension offices, many of which have developed commercially available varieties of food or ornamental crops.

These universities were established for the public good, are still funded by public tax dollars, and are, in principle, beholden to the public trust. These days, many of the land grant institutions receive large endowments from private sector companies, especially large agribusiness corporations like Monsanto, Cargill, Syngenta, and DuPont. A 2012 report from Food and Water Watch found that nearly 25 percent of funding for agricultural research at public institutions comes from private companies.[17] These companies also fund the development of new campus infrastructure, provide money for endowed professorships, and enlist industry representatives to sit on research committees that decide on departmental funding and project proposals. At New Mexico State University, Dow Chemical, Monsanto, and BASF provided herbicide and a series of small grants to help with its efforts to educate landowners about invasive plant species and how to manage them.[18]

Research into the effectiveness of alternative, nonchemical approaches to ecological restoration is lacking in the scientific literature, presumably because it is difficult to procure adequate financial support for these types of studies. Researching invasive species management strategies such as the reintroduction of low-intensity burning or intensive rotational grazing might not benefit corporate interests, but it is exactly what is needed in order to investigate alternatives to herbicide-based restoration.

The collusion between publicly funded land grant universities and pesticide manufacturers is one disturbing trend in the restoration industry. The collusion between federal and state policy makers and pesticide manufacturers is another. Yet another is the relationship between trusted nonprofit conservation organizations and these same businesses.

Rachel Carson dedicated a chapter of *Silent Spring* to denounce the use of organophosphate herbicides (which include glyphosate), which The Nature Conservancy now recommends. In an ironic twist of fate, Carson, who died an untimely death from breast cancer, dedicated one third of her estate to The Nature Conservancy.[19] Today, The Nature Conservancy's website names Monsanto, the manufacturer of DDT, Agent Orange, and other agricultural chemicals whose use Carson opposed, as a corporate partner.[20] Other corporate partners include Dow Chemical, Cargill, Pepsi, Coca-Cola, and multinational mining company Rio Tinto. Companies like Coca-Cola and Cargill that derive profit from converting biodiverse prairies and rainforests into commodity crops like corn for corn syrup and soy for animal feed make strange bedfellows for an organization that's revered for protecting ecosystems from development "for nature's sake." Monsanto donated 120 gallons of Roundup to the Florida Keys branch of The Nature Conservancy for its work on invasive species removal.[21] The Nature Conservancy of Hawaii has received over $130,000 from the Monsanto Fund since 2006 for watershed restoration efforts in Molokai.[22]

Other nonprofit restoration groups, including the National Audubon Society[23] and Ducks Unlimited,[24] also have close ties to pesticide manufacturers, and these companies provide funding or donate herbicides for restoration projects taken on by smaller, more local groups. Southeastern Louisiana University's Turtle Cove Environmental Research Station received a $4,500 grant from the Monsanto Fund to help support wetland restoration programs.[25]

Resource management agencies including the Bureau of Land Management, state Departments of Land and Natural Resources, the US Fish and Wildlife Service, the US Army Corps of Engineers, watershed councils, soil and water conservation districts, and many other state, federal, county, and city agencies use herbicides to manage invasive species in public places like city and county parks, lakes and reservoirs, roadsides, and highway and railroad rights of way.

Maybe the most troubling consequence of pesticide manufacturers' market expansion is the impact it has had on public perception, as well as the perception of land managers and conservation professionals. Today, as agencies and conservation groups grapple with best practices for achieving their land management goals, many view herbicides as regrettable but necessary parts of restoring an ecosystem or protecting native species or property values from the apparent threats posed by invasive species. Although most restoration practitioners take great care to minimize the use of herbicides, the spread and continued proliferation of invasive species means that these chemicals are being used in more and more contexts, with no end in sight.

The Bureau of Land Management estimates that novel organisms invade 4,600 acres of federally managed land in western states every day.[26] Faced with rapidly increasing and seemingly intractable populations of invading plants and animals, many land managers have turned to the use of herbicides and other pesticides in attempts to control their proliferation. Herbicides approved for aquatic use like 2,4-D are applied to invasive plants like Eurasian water milfoil and hydrilla in public waterways near domestic water intakes and recreation areas. Glyphosate and imazapyr-based herbicides are commonly used to control invasive species from dandelions in school lawns to Japanese knotweed and Himalayan blackberries in parks, forests, rangelands, and riparian corridors. In Hawaii, ecologists with The Nature Conservancy load herbicide-filled pellets into modified paintball guns to target Australian tree ferns from hovering helicopters. In Nebraska, amphibious vehicles mounted with spray nozzles patrol the Niobrara River to eradicate purple loosestrife growing along its banks.

Invasive species like Palmer amaranth and leafy spurge also proliferate in agricultural operations. In 2007, farmers reported the application of 442 million pounds of herbicide active ingredients in the quest to control populations of invasive species. In the same year, homeowners and non-agricultural land managers, including conservation groups involved with invasive species management, reported the application of eighty-nine million pounds of herbicide active ingredients.[27] Pesticide manufacturers like Monsanto, Syngenta, Bayer CropScience, Dow AgroSciences, and BASF reaped $5 billion in herbicide sales in the United States in 2007 alone. Worldwide sales of herbicide topped $15 billion that same year. And although herbicides like glyphosate and imazapyr are marketed by their manufacturers as nontoxic, scrutiny of the regulatory apparatus and available research suggests they may be otherwise.

The Precautionary Principle

In the United States, the Environmental Protection Agency (EPA) is responsible for regulating and registering pesticides for use through the Federal Insecticide, Fungicide, and Rodenticide Act (FIFRA). This Act was first passed in 1947, and today all pesticides used in the United States are assessed according to whether or not they cause "unreasonable" adverse effects on people or the environment. According to their definition, the EPA balances "any unreasonable risk to man or the environment" of using a pesticide against the "economic, social, and environmental costs and benefits of the use of any pesticide."[28] This means that although there may be risk in using a certain chemical, it's evaluated in the context of potential economic losses engendered by not using it.

By contrast, in Europe, regulation and registration of chemicals is based on the precautionary principle, meaning that if there is risk of toxicity, the burden of proof falls to those who wish to distribute it. If this principle were applied to the regulation and distribution of pesticides sold in the United States, manufacturers would have to prove that they were not harmful, rather than performing tests that simply determine the level at which they are toxic.

There are a few different ways that the EPA enforces FIFRA. First, the EPA sets standards for allowable levels of chemicals that can "safely" be present in water supplies by establishing maximum contaminant levels (MCLs) for each compound. The MCL for atrazine in drinking water is 3 parts per billion (ppb), meaning that according to the EPA, drinking water supplies can contain 3 ppb atrazine, and people and animals who drink, bathe, cook, and clean with the water will not be at risk for congestion of the heart, lung, and kidneys; low blood pressure; weight loss; muscle spasms; damage to adrenal glands; retinal and muscular degeneration; and cancer.[29] The EPA also sets tolerances for "safe" amounts of pesticide residues on food and forage crops – to date, tolerance levels have been evaluated for nearly 600 pesticides' active ingredients.[30]

The EPA has established that each of these chemicals is safe at some low level, and it regulates products containing them to ensure exposure is at or below these levels during the course of normal application. Of course, determining that it is "safe" to drink 3 ppb of atrazine does not mean that anyone should drink atrazine – ever, let alone every day of their lives. Establishing baselines of "safety" allows for widespread use of these as long as they are dispersed widely enough in the ecosystem to diminish their concentrations. This approach assumes that exposure to potential toxins is an inevitable and necessary part of economic health. It's biased toward the industries that benefit from promoting this belief, and it ensures that these products can be kept on the market by imposing a regulatory framework for their continued use rather than working from a baseline of severely limiting or prohibiting their application.

This approach also makes real-life exposure to multiple pesticides or other chemicals extremely hard to establish. A person may drink or bathe in water that contains atrazine and may consume any number of foods that also contain atrazine. In the course of doing so, he or she could easily exceed the "safe" level of 3 ppb for atrazine. Wheat grain can also contain 2 parts per million (ppm) dicamba, 3 ppm clopyralid, 8 ppm malathion, 2 ppm 2,4-D, as well as 0.1 ppm atrazine.[31] So, if a person consumes conventionally raised wheat, he or she is potentially ingesting dicamba, clopyralid, malathion, 2,4-D, atrazine, and any number of the sixty insecticides, herbicides, fungicides, or other pesticides whose residues may legally be present on the grain.

Though the EPA amended FIFRA in 1996 to consider cumulative pesticide exposures, the interactive and potentially synergistic effects of different types of pesticides have yet to be rigorously investigated. To date, the agency has conducted studies on five classes of pesticides, including organophosphates, triazines, and pyrethrins – but only tested interactions of pesticides of similar types.[32] That is, the EPA looked at potential synergistic effects of different types of organophosphates, but not at the interactions between organophosphates and triazines or any other class of pesticides.

All pesticides registered by the EPA must also undergo a series of toxicity tests, most of which are performed not by the agency, but by the manufacturer of the pesticide. This approach raises important concerns related to potential conflicts of interest, as pesticide manufacturers have a vested interest in designing studies that show that their products are safe.

In 2005, the US Government Accountability Office warned that the EPA lacks safeguards to "evaluate or manage potential conflicts of interest," as it received funding from the industry-sponsored American Chemistry Council to research the health effects of chemical exposures.[33] In an environment where private interests are increasingly represented in public institutions, this trend should be taken quite seriously, especially when it comes to evaluating the potential toxicity of widely applied chemicals.

And although toxicity testing is technically required before a pesticide can be sold and applied in the United States, lapses in the pesticide registration process highlight serious concerns about the rigor of the process. The Natural Resources Defense Council (NRDC) found that 11,000 of the 16,000 plus pesticide products registered for sale in the United States fall under the EPA's "Conditional Registration" program, which waives the requirements for submitting the required toxicity data while the manufacturer assembles the information.[34] The NRDC investigation also found that the EPA does not have a system in place to track whether the data required for full registration – such as the results of toxicological testing – led to any changes in the registration process or even whether it had been submitted in the first place.

When the full complement of tests is complete, manufacturers must assemble data that measure the acute effects of oral, dermal, and inhalation exposure, eye and skin irritation and sensitization, and neurotoxicity. Potential subchronic effects resulting from repeated exposures are measured for oral, dermal, and inhalation exposures. The potential results of chronic exposure, including carcinogenicity, mutagenicity, developmental and reproductive effects, and endocrine disruption are also analyzed as part of the battery of required toxicity tests.

Toxicity is determined by testing the LD50, or median lethal dose, both through skin contact and ingestion. The LD50 of a substance represents the amount that will cause half of the population of a tested organism to die through the particular route of exposure. The oral and dermal LD50 for glyphosate is greater than 5,000 milligrams per kilogram (mg/kg) of body weight for rats (around half a teaspoon), meaning that 50 percent of the rats tested died after ingesting this amount of the herbicide. Following the logic of this approach to toxicity testing, an average-sized person would have to drink more than a cup of glyphosate before they died. Since it is unlikely (but not inconceivable) that a mammal will come in contact with this amount of the herbicide during the course of normal applications, an herbicide that contains glyphosate like Roundup is considered "practically nontoxic," according to the manufacturer.[35]

There are several problems with this approach to measuring toxicity. First, glyphosate is toxic – it will kill a rat or a person if they ingest enough of it. Also, testing for toxicity based on how much an organism was exposed to before it died does not give a full picture of the chemical's potential effects.

Toxicity testing also fails to account for the effects of repeated, long-term exposure, for both the target and nontarget organisms, including the people who apply the chemicals. Though the EPA requires testing for potential effects associated with repeated pesticide exposures, these tests are performed on rats or rabbits over the course of months, which doesn't provide an adequate picture of how a lifetime of pesticide exposure may affect human health. A recent study found a positive correlation between B-cell lymphoma and occupational exposure to phenoxy herbicides, like 2,4-D, and organophosphate herbicides, including Roundup.[36] The authors found that farmworkers tend to have low mortality but high incidences of cancer, especially non-Hodgkin's lymphoma and that exposure to agricultural pesticides is often associated with significant sublethal impacts.

Tests for potentially toxic ecological effects are also required as part of the registration process and are usually carried out on rats, crayfish, rainbow trout, honeybees, earthworms, water fleas, and bobwhite quail as surrogates for determining effects on the major classes of animals (at least mammals, crustaceans, birds, insects, and terrestrial and aquatic invertebrates). Toxicity testing on reptiles and amphibians – well known for their sensitivity to chemicals because of their porous skin – is not required. Instead, the EPA allows data from tests on surrogate animals – rainbow trout or sunfish representing juvenile amphibians, and bobwhite quail or mallard ducks standing in for adult amphibians and reptiles – to determine toxicity to amphibians and reptiles.[37]

Planetary ecosystems contain more than rats, crayfish, rainbow trout, honeybees, earthworms, and bobwhite quail. Though these tests appear to cover the general picture of animal life on Earth, such broadly unspecific tests are probably not demonstrating the whole picture of potential toxic effects. This approach to toxicity testing uses only single classes of organisms to represent effects on the entire ecosystem. Rainbow trout, for example, are used to determine toxicity to aquatic organisms. According to Dr. Meriel Watts of the Pesticide Action Network Aotearoa New Zealand, "the healthy functioning of aquatic ecosystems depends on a wide variety of organisms including microorganisms, algae, and amphibians, effects on which are seldom if ever measured for [herbicide] registration purposes. The significant detrimental impact of glyphosate/Roundup on these less-regarded species is of far greater ecological importance than the singular measure of acute toxicity to fish."[38]

Toxicity testing should also require the examination of the effects upon *all* organisms that could come in contact with the herbicide, including those that are underrepresented in scientific literature or public perception. Bacteria, fungi, and other microorganisms are as or more important to ecological health as mammals, for example, even though it's the mammals that tend to get the spotlight. Some biologists refer to this as a duality between "charismatic megafauna" and "despauperate microflora" – but size, visibility, and cuteness do not dictate an organism's value to the health of an ecosystem.

Such extensive testing for toxicity would be expensive and likely impossible for pesticide manufacturers to undertake since there are uncountable species, especially microorganisms, that haven't yet been named or identified. In June 2012, the EPA released "harmonized testing guidelines" for pesticide registration that include more far-reaching and comprehensive ecological analysis. Some of the new tests include assessments of how a pesticide affects soil microbial communities, including populations of cyanobacteria and nitrogen-fixing bacteria, as well as testing for how pesticide residues on foliage may affect honeybees.[39] However, the EPA currently recommends rather than requires pesticide registrants to use these new guidelines, and pesticides already on the market have not been analyzed according to these more comprehensive standards.

Active and Inert Ingredients

Pesticides also contain both active and so-called inactive or inert ingredients. Many formulations contain more than one active ingredient. For example, the herbicide Surge includes four herbicide active ingredients, including sulfentrazone, 2,4-D, MCPP (mecoprop), and dicamba, a potent mixture designed for killing broadleaf plants and recommended for use on such invasive species as yellow nut sedge, leafy spurge, and Japanese knotweed. And though Surge contains four different products, the plant-killing chemicals make up only 30 percent of the formulation. The remaining 70 percent is made up of "inert" ingredients, including surfactants, solvents, emulsifiers, wetting agents, polymers, and adjuvants. These compounds help the herbicide stick to the fuzzy, hairy, or waxy outer surface of a plant's leaves and help to break the surface tension of water when used in aquatic environments. They are designed to make the active ingredient(s) effective despite variable weather, such as wind, rain, dew, and fog, and stay on the plant for weeks at a time. The type and exact amount of these "inert" ingredients in any formulation are trade secrets and are not listed on the herbicide labels beyond the percentages of active and inert ingredients like the 30 percent active: 70 percent inert ratio shown on the label for Surge.

The active: inert ingredient ratio can vary, but most registered herbicide formulations contain between 2 percent and 50 percent active ingredients. The EPA reports that 531 million pounds of herbicide active ingredients were applied in the United States in 2007, the last year that data are available.[40] If 531 million pounds of active ingredients were applied, then it's reasonable to estimate that up to 10 billion pounds of these unregulated "inert" ingredients were applied to soils and waterways that same year. These numbers are just from the United States alone and are based on reported applications of herbicides. Actual numbers have likely increased since 2007, and actual application rates are almost certainly much higher.

Every two years, Southern Illinois University publishes *Compendium of Herbicide Adjuvants*, a list of known herbicide adjuvants. Whereas the first edition, published in 1992, cited 72 adjuvants used by twenty-two companies, the 2012 edition cites 687 adjuvants made by thirty-eight companies – suggesting rapid growth in the industry of (plant) death. Market research on the proliferation of adjuvants reveals that sales of adjuvants are expected to reach over $3 billion in 2019, making their production and use one of the fastest-growing segments of the agricultural technology industry.[41]

Some herbicide formulations, like Rodeo, Accord, and Aquamaster, are recommended for use in restoration contexts, especially in aquatic ecosystems, because they do not contain surfactants, which interfere with respiration in fish and frogs and therefore show "favorable toxicological characteristics" – meaning they aren't known to suffocate fish and amphibians.[42] If an herbicide is purchased as a single active ingredient (like glyphosate) rather than as a named, trademarked formulation (like Roundup), the manufacturer's label will often provide recommendations for the addition of types of adjuvants that will enhance the success of the herbicide. For example, the manufacturer of imazapyr recommends that the addition of crop oil (petroleum), a nonionic surfactant, a silicone-based surfactant, or fertilizer–surfactant blends could be used to enhance the effectiveness of the herbicide itself.[43]

Given the wide array of available adjuvants, and the many different applications for which the herbicide formulation could be used, it can be difficult to assess which one is best (and least likely to negatively impact the surrounding ecosystem). The specific recommendations for adjuvants vary according to whether the target plants are aquatic or terrestrial; the weather conditions on the day of application; the pH of the water used to mix the herbicide; the life stage, density, and growth habit of the target species; as well as the method of application.[44] Compounding the complexity of making the best decisions for the selection of adjuvants is their lack of regulatory scrutiny, which makes it difficult to determine best practices regarding their use.

Though the EPA maintains a list of all ingredients contained in pesticide products, only the so-called "active" ingredients appear on the labels; the applicators, the consumers, and the general public have no means by which to know the full spectrum of ingredients contained in a particular pesticide. According to federal law, pesticide ingredients that don't harm the target pests are not considered active ingredients and don't have to be shown as part of the pesticide label.[45] This is ostensibly because the particular products a pesticide manufacturer uses to make its product stand out in the market are trade secrets and cannot be made publicly available for fear of competition from other manufacturers. In 1999, only 10 percent of more than one hundred commonly available pesticide products sold in retail stores in New York identified *any* of the inert ingredients on the label. None of these labels identified *all* of the inert ingredients in the products.[46] This trend is disturbing, because in common usage, the word *inert* is used to describe something benign or biologically inactive, a definition that does not fit many of the "inert" ingredients in pesticides.

A 1998 report by the Northwest Coalition for Alternatives to Pesticides (NCAP) found that 209 inert ingredients are considered hazardous air or water pollutants, 21 have been classified as carcinogens, and 127 as occupational hazards by agencies including the International Agency for Research on Cancer, the National Toxicology Program, and the State of California. In addition, almost 400 "inert" ingredients are or have been used as active ingredients in pesticides, they just aren't listed as such in other formulations because they are supposedly not targeting the intended pest.[47]

The type of testing required for registration of active ingredients, though limited in scope, is more substantial than that required for the so-called inert ingredients or the formulations in which they are found. Caroline Cox of NCAP and Michael Surgan, chief scientist for the attorney general of New York, sum up the inconsistencies of toxicity testing requirements between active ingredients and pesticide formulations available on the market:

> Of the 20 toxicologic tests required to register a pesticide in the United States, only 7 short-term acute toxicity tests use the pesticide formulation; the rest are done with only the active ingredient. The medium- and long-term toxicity tests that explore end points of significant concern (cancer, reproductive problems, and genetic damage, for example) are conducted with the active ingredient alone. The requirements for other types of tests are similar. Only half of the required (or conditionally required) tests of environmental fate use the formulated product, as do only a quarter of the tests for effects on wildlife and non-target plants.[48]

Thus, toxicity tests on products like Roundup, are largely based on examining the active ingredient glyphosate rather than the full array of ingredients contained in Roundup. How widely used herbicides – not just their active ingredients – affect the health of birds, amphibians, insects, invertebrates, soil biota, and other mammals, including humans, that come into contact with them is largely unknown.

Although specific pesticide formulations are considered trade secrets, the identities of some of the more widely used "inert" ingredients are known. Popular surfactants include POEA (polyethyloxylated tallow amine) and paraffin-based crop oils. Polymers like polyacrylamide and polyvinyl (plastics) are added to mitigate drift potential for herbicides sprayed from helicopters or backpacks. Water conditioners such as ammonium sulfate are added to enhance the reactivity of aquatic herbicides where hard water is present, since some herbicides (like glyphosate) are precipitated by alkaline water and rendered ineffective. Silicon-based defoaming agents are also often part of the mix because the surfactants and emulsifiers can produce bubbles that disrupt the effectiveness of the applicator nozzle once mixed with the herbicide active ingredient(s).

The toxicity and ecological impact of these "inert" ingredients warrants investigation, especially since the addition of surfactants, adjuvants, petroleum-based polymers, plastics, water-softening agents, defoamers, and other emulsifiers enhances the efficacy of herbicides by making the active ingredient(s) more biologically available, absorbable, and longer lasting on the plants to which they are applied. A 2007 study found that Roundup, which contains the active ingredient glyphosate in addition to unknown "inert" ingredients, is more toxic to human mononuclear cells – including those critical to healthy immune function found in bone marrow cells – than glyphosate on its own.[49] Another study found that Roundup was particularly toxic to human placental cells, more than glyphosate on its own. In this same study, researchers also found that Roundup reduced the activity of aromatase, the enzyme responsible for estrogen synthesis, at concentrations ten times lower than recommended use in agriculture.[50] Experimental studies suggest that the surfactant POEA, one of the known "inert" ingredients in Roundup, may be at least partially responsible for these findings.[51]

The EPA considers glyphosate "slightly to moderately toxic" to humans when used at the manufacturer's recommended levels. However, land managers rarely use glyphosate alone and instead use glyphosate-based herbicide formulations, like Roundup. When looking at the differences in toxicity between glyphosate and Roundup, French biologists found that combining glyphosate with adjuvants like POEA amplifies the toxicity of the herbicide by changing the permeability of human cells and inducing cell death.[52] Monsanto, the manufacturer of Roundup, denounced these studies; it asserted that they lacked credible methodology and drew conclusions based on laboratory results that vary from real-world conditions. The results at the very least point to a need to identify, test, and regulate the use of the so-called inert ingredients found in widely used pesticide formulations. In 2009, the EPA began the process of figuring out how to disclose information about confidential or proprietary inert ingredients to the general public. In 2014, it removed seventy-two chemicals from the approved inert list, though as of this writing, it has yet to disclose the full spectrum of inert ingredients in registered pesticide formulations.

Independent research on the long-term, nonlethal, and ecological effects of herbicides commonly used in restoration contexts demonstrates that the toxicity tests required for their use and application are limited in scope and leave much to be desired in terms of making sound assessments regarding their safety for widespread use. The Nature Conservancy describes best practices for applying eleven different herbicide active ingredients in its *Weed Control Methods Handbook for Natural Areas*.[53] Three of the most widely used – 2,4-D, glyphosate, and imazapyr – will be discussed here.

2,4-D

2,4-D was originally one of the ingredients in Agent Orange, the herbicide widely dispersed throughout Vietnam, Laos, and Cambodia during the Vietnam War. The other ingredient in Agent Orange, 2,4,5-T, has since been banned because many of the batches were contaminated with dioxin as part of its manufacture, leading to widespread incidences of cancer and birth defects both in the populations of Southeast Asia and in soldiers returning from the war.[54] Agent Orange was also used to clear "brush" from industrial forestry operations in California, Oregon, Washington, and throughout eastern Canada prior to its wartime applications.

Though 2,4,5-T is rarely used in the United States these days, 2,4-D remains a popular herbicide for home use, as well as in agriculture and restoration. As a broadleaf-specific herbicide, it does not kill grasses, conifers, reeds, or rushes, so it is widely used in the cultivation of nonorganic grains like wheat, corn, and oats. Around forty-six million pounds of 2,4-D are applied in the United States annually, with approximately eleven million pounds applied on rangeland and pasture to manage invasive species and eight million pounds used by homeowners on their lawns largely to remove dandelion and clover.[55]

2,4-D is sold under the trade names Aqua-Kleen, Plantgard, and Restore, among others. In restoration, 2,4-D is used to manage invasive species like giant hogweed, Russian thistle, and hoary cress. It is also commonly used in aquatic ecosystems on Eurasian water milfoil and water hyacinth. 2,4-D is considered relatively benign because it is a synthetic version of auxin, a natural plant hormone that regulates growth and development. When applied to a plant, 2,4-D acts like auxin, but it persists, leading to rapid and abnormal cellular growth that eventually kills the plant.[56] Its propensity to encourage cell proliferation has led to ongoing controversy as to whether 2,4-D is carcinogenic, as cancerous cells demonstrate similar tendencies toward rapid and uncontrolled cellular growth resulting in tumors or lesions.

Research shows that farmworkers who have applied 2,4-D and other phenoxy herbicides are at greater risk for non-Hodgkins lymphoma[57] and that 2,4-D applicators experienced rapid and repeated division of blood cells following exposure to the chemical.[58] In addition, though the dioxin-containing 2,4,5-T is no longer registered for use in the United States, the EPA found that two of every eight samples of 2,4-D were contaminated with 2,3,7,8-TCDD, another type of dioxin, as well as other dioxins.[59] This study was undertaken as part of an analysis of how commonly used agricultural chemicals including 2,4-D would affect federally listed threatened species in California, including the red-legged frog and Alameda whipsnake. It found that all uses of 2,4-D (outside of citrus and potatoes) are likely to adversely affect populations of these animals.[60] The National Institute of Occupational Safety and Health advises that 2,3,7,8-TCDD can cause reproductive problems and be carcinogenic and mutagenic in small doses.[61]

The EPA categorizes 2,4-D as "not classifiable" to human carcinogenicity,[62] although it also states that 2,4-D "has not undergone a complete evaluation and determination . . . for evidence of human carcinogenic potential."[63] Though the phrase "not classifiable" may be taken to infer that 2,4-D is not a carcinogen, it really means that there is not enough evidence to tell one way or another. Much of the data used to make these determinations is several decades old, and many of the studies looked at all phenoxy herbicides (including the now banned 2,4,5-T) rather than at 2,4-D specifically.[64] The EPA does acknowledge that 2,4-D is toxic to the eye, thyroid, kidney, adrenals, and ovaries/testes, but only when at high levels of long-term exposure.[65] It also notes that 2,4-D exposure poses concern for neurotoxicity, which further studies have corroborated by demonstrating reduction in brain size and disrupted connections between nerve cells and the brain.[66]

Glyphosate

The development of glyphosate came on the heels of the Vietnam War, when Agent Orange was decried for its long-lasting toxic effects on people and its persistence in the environment. Glyphosate was hailed as a safe alternative and was recommended for use throughout agriculture and forestry. Glyphosate's application as a tool to manage invasive species grew out of the nascent Cal-IPC's use of the herbicide (under the technical advisement of Monsanto employee Nelroy Jackson) to manage giant reed in southern California in the early 1990s. It remains one of the most popular active ingredients in the herbicides used in agriculture and restoration. Glyphosate-based herbicides are largely considered safe for wildlife, as industry tests have shown low to moderate toxicity for all tested organisms. The oral and dermal LD50 is relatively high for glyphosate, at greater than 5,000 mg/kg. Most organisms would have to consume more than a teaspoon of straight glyphosate before they died, which is possible but unlikely when used according to the manufacturer's application recommendations. But of course, in the real world, the degree and duration of exposure that an organism might experience can still vary widely, unlike the closely controlled simulations in laboratory tests.

Glyphosate was originally patented in 1964 as a descaling agent to be used in the removal of mineral deposits from dishwashers, heating vents, and other appliances where evaporating water causes mineral accretion. The herbicidal properties of glyphosate were accidentally discovered when it killed the grass growing beneath an air conditioner being cleaned with the substance. Monsanto patented the product for use as an herbicide in 1974, but the compound still exhibits the chelating properties for which it was originally patented. Glyphosate chelates zinc, copper, manganese, iron, calcium, and magnesium, so when it comes in contact with these elements, they bind with the molecule and become biologically unavailable.[67] When these vital elements are chemically "locked up" by the glyphosate molecule and the shikimate pathway – a finely tuned series of enzyme interactions used by plants, fungi, and bacteria to synthesize amino acids – is interrupted, the immune system of the plants to which it is applied is compromised. In other words, glyphosate does not actually kill plants per se; it renders them vulnerable to soil-borne diseases by destroying their immune systems.[68] Mineral deficiencies make their way through the food web: People and livestock who are exposed to glyphosate residues in food are prone to (potentially pathological) mineral deficiencies, and birds, fish, insects, and other mammals are also affected as minerals are bound up in the soils of sprayed forests and wetlands.

Because glyphosate interferes with the shikimate pathway, which is the primary means by which many bacteria access energy to proliferate, it was patented as an antimicrobial in 2010.[69] One genus of bacteria named in the patent as targets of glyphosate are *Pseudomonas*. Some species of this genus cause plant diseases, but others assist with phosphate mobilization in soil, and others aid in the suppression of the soil-borne plant pathogen *Fusarium*. Glyphosate's effect on soil microorganisms is not well-researched, but should not be overlooked, since microbes form the basis of the terrestrial food web.

Although glyphosate has been patented for its use as an antimicrobial, it's largely considered nontoxic to organisms lacking the shikimate pathway – insects, birds, amphibians, and mammals. However, the human body alone is home to an estimated 1,500 species of bacteria numbering upwards of one hundred trillion individual organisms. In fact, only 10 percent of the human body is made up of human cells; the rest are bacteria.[70] These bacteria play important roles in developing and maintaining the immune system.[71] So although mammals, birds, amphibians, and insects may not exhibit immediate signs of toxicity from glyphosate because they as individual organisms lack the shikimate pathway, the fact that glyphosate acts on bacteria that live in and form the basis of animal immunity has profound implications. A 2013 study found that beneficial bacteria in chicken intestinal tracts were moderately to highly susceptible to – meaning destroyed by – glyphosate,[72] and another published the same year showed that bacteria in the genus *Enterococcus* – which ward off pathogenic bacteria like botulism – were also negatively influenced by glyphosate.[73] As of this writing, no similar tests have been performed on bacteria unique to the human intestinal tract.

Beyond the concerning and untested potential health effects of glyphosate in animals, in restoration contexts, plants are often sprayed and left to decompose in place, and so the glyphosate will remain present in the ecosystem long after the "eradication" takes place. Glyphosate-based herbicides are promoted as ecologically benign because they are not mobile in soils and do not readily leach into groundwater or travel far beyond the area of application. However, again because of its powerful chelation ability, glyphosate binds tightly with minerals in the soil and does not easily biodegrade. The mineral-chelating properties of glyphosate also affect the soil over the long term, changing its chemistry, and perhaps even affecting the type of plants that are able to grow there.

If the goal of restoration is to return a site to its "original" state – and this is the stated goal of many restoration projects – then glyphosate-treated soil with artificially limited zinc, copper, iron, calcium, manganese, and magnesium cannot possibly approximate historical soil characteristics. Invasive species are targeted by herbicides in restoration because native species did not evolve with their competitive pressures, and many invasives are accused of altering soil characteristics where they are found. Native plants did not evolve in soils where primary micronutrients are strongly bonded and biologically unavailable because of synthesized molecules. Native species rely on a vast array of functional relationships among soil microorganisms and fungi whose populations are depleted or destroyed by the use of glyphosate-based herbicides. If the goal of ecological restoration is to approximate precontact ecosystem conditions, then herbicides like glyphosate cannot be a part of the management strategy.

Imazapyr

Imazapyr-based herbicides have become more popular in restoration settings, especially for aquatic applications. Imazapyr-based formulations such as Habitat are routinely applied on salt cedar, common reed, spartina, and water hyacinth. The material safety data sheet for Habitat warns against applying it within half a mile of potable water sources, including rivers, streams, reservoirs, and lakes and then goes on to describe how to use it in these same ecosystems. A forty-eight-hour waiting period after application is recommended before turning the water intakes back on, a meaningless waiting period for organisms that live where Habitat has been applied.

According to the manufacturer, BASF, Habitat and other imazapyr-based herbicides have a "high probability" of not being toxic to aquatic and terrestrial life (except, of course, the plants that aquatic and terrestrial life depend on for food). Laboratory testing for these claims is limited, as with all registered pesticides, measuring toxicity from single exposures of relatively short (up to ninety-six hours) duration. Long-term exposure to the compound is another matter (though this important research is not necessary for labeling or application). One study found that female rabbits fed imazapyr over the course of twelve days developed stomach ulcers and intestinal lesions.[74] Two year-long studies on rats found fluid accumulation in lungs, increased incidence of kidney and thyroid cysts, increased brain congestion, as well as pooling of blood in the liver and spleen – none of which were considered significant toxicological concerns by the EPA.[75]

Imazapyr, like glyphosate, is resistant to decomposition in soil and has been found in soils more than a year after application.[76] Its herbicidal effects are still active during this time, which is considered beneficial in restoration contexts to suppress continued growth of the targeted species. However, the long-lasting nature of imazapyr means that its acute and chronic toxicological effects on the organisms with which it comes in contact also continue to act well past the initial application. Imazapyr is also highly mobile in both water and soil and has been found nearly 10 feet below the soil surface to which it was applied.[77] All of these concerns point to the fact that imazapyr is not any more "safe" than other herbicides commonly used in restoration.

All things considered, though most common herbicides are considered benign or nontoxic, there is significant evidence pointing to the opposite conclusion. Important questions regarding their potential toxicity to a wide variety of organisms, unknown interactions with soil microorganisms, residual effects on desired vegetation, and synergistic effects of their mixtures remain unexplored. Research into the toxicological profiles of the so-called inert ingredients in commercial herbicide formulations is also needed. Manufacturing companies should be required to disclose all ingredients to consumers and regulatory agencies; corporations with vested interests in particular outcomes, such as "safety," should not be considered valid sources of or collaborators in scientific findings; and scientists need to disclose their funding sources in all published articles so people can judge for themselves if the results or nature of a experiment may be biased toward particular outcomes.

Lasting Legacies

Like imazapyr, other herbicides persist in the soil long after they've done their job killing the vegetation. Even manufacturers state that soil residues are inevitable. Bayer CropScience reports that 30–100 percent of their "crop protection products" can end up in the soil.[78] According to the University of Illinois, most herbicides persist in the soil for one to twelve months, but several common herbicides, like imazapyr and picloram, persist for longer than one year.[79] The length of time that residues can be found also varies depending on the soil type, pH, cation exchange capacity (CEC), and number and type of soil microorganisms present.

The persistence of herbicides in soils raises questions about their potential toxicity for soil microorganisms. The few studies that have been performed on its interactions with soil organisms show that glyphosate decreases the length of fungal hyphae,[80] interferes with nitrogen-fixing bacteria, and increases the incidence of pathogenic soil bacteria.[81] We rely on soil-borne organisms for the breakdown of these herbicides, yet the manufacturers are not required to test the effects on soil microorganisms.

The functional life of an herbicide does not stop with its application. Herbicide active and inactive ingredients are complex molecules that are broken down over time into their components. Some are soluble in water, while others are suspended in soil. Some herbicides, like Roundup, are considered safe because the glyphosate molecule is not detectable in soils after a relatively short amount of time post-application. Research into the persistence of glyphosate shows that it's half-life in soil can vary from between 3 and 130 days. When it does break down, the product of its microbial decomposition – aminomethylphosphonic acid (AMPA) – lasts even longer. The end product of the decomposition of glyphosate is carbon dioxide. The USDA found that 55 percent of glyphosate applied is released into the atmosphere as carbon dioxide within four weeks of application. Given that we're facing potentially catastrophic climate change in the years ahead, we need to scrutinize the widespread use of glyphosate on these grounds as well.[82]

Glyphosate is considered environmentally benign because it cannot be found in soil after three months in average conditions. However, AMPA, its by-product, has been shown to last in the soil for up to 958 days, and AMPA adsorbs especially well into soils with high organic matter.[83] This molecule has a similar toxicological profile to glyphosate, yet has never been adequately tested for toxicity. The breakdown product of imazapyr, quinolinic acid, is considered neurotoxic, causing nerve lesions and symptoms similar to Huntington's disease in humans.[84] Little is known about the effects of herbicide surfactants and adjuvants in soil or their breakdown products.

Herbicide Use in Practice

Many commercial herbicide formulations consist of more than one herbicide active ingredient, like Surge described above, which contains four different compounds plus the unlisted "inert" ingredients. Though each individual active ingredient has been tested for toxicological characteristics, potential synergistic effects from these mixtures are unknown. Each of these products comes with a list of manufacturer's recommendations for timing and ideal conditions for application. However, the broad array of products and their specific recommendations could make it difficult for a land manager to keep all the rules and regulations straight. Although several commercial formulations that include more than one active ingredient are available on the market, it can be tempting for land managers to formulate their own untested and unapproved mixtures in attempts to increase the potency or longevity of the application.

The ideal conditions for application – typically warm, dry, and windless conditions – are also difficult to pin down in a constantly changing world. Wind and rain can lead to drift or runoff. Some herbicides, like picloram, are not licensed for aquatic use because of toxicity concerns and long-lasting residues, but what happens if it rains six hours after it is applied? If the herbicide is so durable in the environment, it is likely that it will enter the aquatic environment at some point, since all terrestrial and aquatic ecosystems are ultimately connected. Even the most careful application of herbicide to the surface of a plant will at some point make its way into the watershed, either directly through inadvertent contact or as the plant decomposes and soil microorganisms transport the substance throughout the subsurface food web.

Although specific recommendations are available for each herbicide in a variety of contexts, there is no guarantee that the applicator will follow these instructions. Unlike some farmers, who have experience and perhaps even training around the use of chemicals in their operations, those working in restoration may have little education around their proper application, other than what they receive through word-of-mouth instructions from colleagues. A recent demonstration at the University of California

Cooperative Extension in southern California showed that herbicide applicators working in restoration were applying significantly greater amounts of herbicide than recommended by the manufacturer and that they were mostly unaware of their level of use.[85] The study found that people routinely applied two to fourteen times the recommended amount of herbicide active ingredient during field testing and had no idea they were doing so until researchers measured the amount of herbicide sprayed into a bucket for the same amount of time it took the applicators to cover a test plot. These findings were so shocking to the researchers that they held trainings to teach correct use. The bottom line is that it's incredibly difficult, perhaps impossible, to know how much exposure humans and ecosystems actually have to these herbicides in practice.

Nontarget Effects and the Pesticide Treadmill

Broad-spectrum herbicides like glyphosate and imazapyr are designed to kill all plants, even native plants. Use of herbicides as a restoration tool sets back the native plant populations they are designed to "protect." Repeated applications of herbicide are normal restoration protocol, which further diminishes the ability of native flora to establish or recover. One study found that a single herbicide application in an attempt to remove spotted knapweed reduced native arrowleaf balsamroot populations by more than 40 percent and negatively affected the seed production and density of the surviving plants.[86] Such drastic reduction in the desired native forbs leads to more open ground – the condition spotted knapweed prefers.

This cascade of events is similar to the pesticide treadmill of conventional agriculture. When a farmer applies insecticides, it affects more than the target species. Insecticides kill the beneficial predatory insects as well as those that eat crops. A single application is unlikely to kill all of the target organisms since they may be present in different life stages (eggs, larvae, adult, etc.) or out of the range of the pesticide applicator nozzle. Surviving individuals are more likely to have some level of resistance to the chemical applied – perhaps they were paralyzed or stunned rather than killed immediately – and so are able to reproduce in an environment free of the predators that were killed by the pesticide. In order to curb unchecked growth of greater and greater pest populations, conventional farmers apply more and different pesticides in a difficult-to-break cycle.

Plants can also develop resistance to herbicides, which is becoming a significant issue in agricultural operations that use genetically modified crops, like Roundup Ready corn or soybeans, which are designed to survive repeated herbicide applications. For the weeds, frequent use of broad-spectrum herbicides constitutes intense selective events. From an evolutionary perspective, this practice pressures the weeds to develop novel mechanisms of survival in the face of chemical assault.

Herbicide-resistant horseweed has been shown to shuttle 85 percent of the glyphosate applied into its vacuoles, where it cannot harm the cell and can later be disposed of. The remaining 15 percent moves to the leaf tips, which can die back and leave the plant with enough energy to continue growing from its base.[87] Herbicide resistance frustrates farmers and industry leaders, prompting the development of crops resistant to 2,4-D and other herbicides in the attempt to evade the tendency of increasing herbicide resistance of common weeds. This will simply intensify the pesticide treadmill with yet more widespread application of a more toxic chemical.

Although herbicides are not applied at the same rates in restoration as in agriculture, the practice does bring up questions of how long their use in this capacity will be possible. Will commonly sprayed invasive species like purple loosestrife, Japanese knotweed, and Himalayan blackberry develop herbicide resistance? If this is the case, the next steps for chemical-based restoration are unclear. This is one of the many reasons that it is necessary to evaluate invasive species and restoration with a holistic approach – one that does not rely on the application of increasingly ineffective and dangerous chemicals.

Nonchemical Control

Although herbicide use is widespread in the context of invasive species management, most people in the restoration community consider herbicides a less than ideal option because of the known risks associated with their use. Land managers sometimes use mechanical methods like mowing, tilling, burning, and digging either alone or in concert with an herbicide in order to clear an invaded area, so that they end up using less herbicide to achieve their goals. But these methods aren't useful in all restoration projects since they tend to be unspecific: When a rototiller tills up soil, it tills everything, not just the target species. Sensitive or endangered native plants or animals might be harmed by these methods. They also might not be economically feasible when the invaded area is extensive. Mechanical control is a useful tool in nonchemical invasive species management, but people still typically use it within the framework of eradicating the invasive species quickly and completely – a paradigm that fails to account for the whole system and thus is not ultimately effective.

Land managers also control certain invasive species biologically, by introducing herbivores that eat invasive plants, predators that prey on invasive animals, or diseases that might kill either. While biological control demonstrates a more holistic approach to managing invasive species, by considering them as part of a greater ecological system, it also introduces new organisms, which can be counterproductive and potentially fraught with the risk of unintended consequences – something especially likely given the complex nature of ecological interactions. The US Department of Agriculture (USDA) regulates the testing and release of biological control agents, most of which undergo over five years of rigorous testing to determine whether they will affect nontarget organisms. The US Fish and Wildlife Service estimates that only one-third of introductions of biological control agents have met with "success," which is defined as partial or local eradication of invasive species.[88]

Some biological control experiments have altered ecosystems in unforeseen ways. The cactus moth larvae, which were introduced to the Caribbean to manage populations of prickly pear cactus, have since moved into Mexico, Florida, and the southwestern United States, decimating native cactus populations. The thistle head weevil was introduced to the United States in 1969 to control the invasive musk thistle, but they also fed on the seeds of rare and endangered native thistles. Some populations of this weevil also carried eggs of the microsporidian fungus nosema in their digestive tracts. The introduction of these beetles inadvertently led to the establishment of this devastating honeybee parasite in the United States. Subsequent research found that fourteen of forty-two species being researched as potential biological control agents were contaminated with nosema or other pathogens.[89]

The introduction of biocontrol species is bound to have interactions throughout the larger food web as well. Many biocontrol agents are herbivorous insects, which defoliate, weaken, and eventually kill or disable reproduction in invasive plants. Such insects are preferred food sources for birds, snakes, amphibians, and rodents. If the insect biocontrol species is flourishing, then other members of the biotic community will flourish as well, such as deer mice that feed on knapweed gall flies. Thanks to the abundance of fly larvae, the population of deer mice, which also feed on native vegetation, can increase two- or threefold.[90]

Where biological control has been successful, where populations of invasive species have declined or disappeared because of their introduction, the ecosystems now free from the target species are still open for invasion. One study found that following successful suppression of invasive St. John's wort through the introduction of seed-eating beetles in Idaho, invasive yellow star thistle and nonnative annual grasses colonized the research sites.[91] The prevalence of yellow star thistle is especially ironic, because the point of controlling the St. John's wort was to improve forage value for cattle (for which St. John's wort is poisonous), and yellow star thistle is even less palatable.[92] Biological control strategies can be effective, but it is impossible to predict all of the factors – and potential unintended consequences – at play in complex ecosystems.

Today, biological control agents are subject to intense scrutiny and years of research to determine their potential ecological effects. This is one of the only areas where the regulatory process in the United States abides by the precautionary principle: Potential biological control agents must be proven relatively safe in their ecological effects before they can be released. The risks that they potentially pose must also be balanced against the apparent risks of managing invading organisms in other ways or not managing them at all.[93] If the herbicides and other chemicals employed in the absence of biological control were subject to the same level of scrutiny, far fewer chemicals would be used in the name of ecological restoration.

A More Rational Approach to Restoration

We must look beyond this paradigm that seeks to eradicate invasive species by any means in order to meet the goals of restoring native plant and animal communities. Restoration projects must design and implement plans that take the full picture of ecological health into consideration. Otherwise, restoration is no better than conventional agriculture – using the same tools (pesticides) and based on the same mindset (necessity to control an apparently recalcitrant nature). The use of herbicides accomplishes a single outcome – the eradication of plants – without taking the larger systemic picture of ecological processes into account.

Ecological restoration can, and should, mean more than spraying plants with toxic substances. If the definition provided by the Society for Ecological Restoration is accurate, then the practice of restoration is the process of assisting with the repair of degraded ecosystems. Such practices should not entail their further degradation. Though Monsanto and other pesticide manufacturers promote the concept that their products can be used to "protect and restore" ecosystems, there are more meaningful, longer-lasting, and mutually beneficial ways that ecosystems can be restored to high levels of diversity and functionality. In order to repair degraded ecosystems, we must learn to think like an ecosystem, to use our intelligence and capacity to shape terrestrial and aquatic environments in ways that enhance their overall diversity and functionality. The restoration industry is caught in the thrall of misguided concepts in desperate need of updating: The presence of invasive species is not necessarily a problem to be solved, but rather an invitation to delve deeply into understanding the complex ecosystem dynamics to which they are intrinsically related.

CHAPTER 2

Getting to the Root of Invasive Species

"It is thus necessary to examine all things according to their essence, to infer from every species such true and well established propositions as may assist us in the solution of metaphysical problems."

— *MAIMONIDES*

Washington's Willapa Bay is the second largest estuary in the United States, protected by a long peninsula from the pounding waves of the Pacific Ocean. It is also known as one of the cleanest estuaries in the world, provides habitat for green sturgeon, smelt, Dungeness crabs, and juvenile salmon, and is an important stopover for birds migrating along the Pacific flyway – one of the largest migration routes in the world. Cultivated marine species including oysters, clams, mussels, and geoducks thrive in the extensive intertidal zones; each year farmers in the bay retrieve three to four million pounds of oysters from 10,000 acres of privately owned tidelands.

Native Olympia oysters once thrived in the bay, until overharvesting – mostly to feed gold miners in the California Gold Rush – decimated their populations by 1900. Noticing the declining native oyster populations, enterprising mariculturists began importing seed oysters from Massachusetts and Connecticut in the late 1800s to bolster oyster production in Willapa Bay. (This continued until the 1920s when they began to favor the Pacific oyster from Japan for intensive cultivation.) To protect the seed oysters, East Coast shippers packed them in crates with grasses that were harvested from the salt marshes where the East Coast oyster grew – grasses which played, and continue to play, a critical role in maintaining the diversity and productivity of the estuarine ecosystem on the East Coast. Both the grasses and the oysters they cradled survived the journey and adapted to their new ecosystems. One of these crate-packing grasses, spartina, took root in Willapa Bay.

For nearly a century, spartina grew mostly in small, isolated patches. But by the mid-1980s, patches of spartina expanded to cover close to 300 acres in the bay, and in the 1990s, its spread increased exponentially, peaking at more than 10,000 acres in 2011. This recent and rapid spread of spartina now threatens Willapa Bay's $30 million oyster industry, which relies on the presence of open mudflats for easy access to the seedbeds for young oysters. Spartina accumulates sediments in its dense root system and has been known to raise the elevation of mudflats by 1–2 centimeters (cm) per year.[1] If left to grow, spartina could transform the prime oyster-producing mudflats into grassy marshes.

Oyster producers first sounded the alarm about the spread of spartina as far back as 1942, but the plants were only found in a few locations at that time, and few oyster farmers were affected. By 1979, however, the Washington State Department of Fish and Wildlife recommended its eradication from all state estuaries.[2] Oddly, only six years earlier, the US Army Corps of Engineers had introduced spartina to stabilize shoreline erosion in the San Francisco Bay. Populations of spartina soon spread in the San Francisco Bay, and today the plant is considered an invasive species throughout the West Coast, where it hybridizes with native species of spartina and proliferates in formerly vegetation-free mudflats. The California Invasive Plant Council (Cal-IPC) warns that spartina alters hydrology and clogs flood control and navigation channels because of its tendency to trap and hold sediments. And many ecologists are concerned that shorebird populations will decline because of spartina's spread since its dense roots support different organisms than those found in open mudflat environments where the birds currently forage.[3]

Starting in 2003, oyster farmers in Willapa Bay joined forces with government agencies including the Department of Fish and Wildlife and restoration organizations such as The Nature Conservancy to implement a massive spartina eradication campaign in an attempt to mitigate the spread and potential ecological consequences of spartina. For the next nine years, these groups applied the imazapyr-based herbicide Habitat to more than 9,000 acres of spartina. Boats, helicopters, backpack tanks, and amphibious tractors sprayed the estuarine grass with herbicides, adjuvants, dyes, and surfactants. Today, the eradication campaign is lauded as a success – only a few acres of spartina remain in Willapa Bay – and similar eradication efforts are underway in Oregon and California estuaries. In 2005, more than 2,000 acres were sprayed with imazapyr- and glyphosate-based herbicides in and around the San Francisco Bay.

Meanwhile on the East Coast, the same species of spartina is prized for its habitat value and ability to stabilize shorelines. Here, spartina is also known as oyster grass because the marshes where it grows along the Gulf Coast, Chesapeake Bay, and Cape Cod were once productive oyster beds, prior to the decimation of regional shellfish populations from industrial development and pollution. From Texas to Maine, spartina forms the basis of a diverse salt marsh food web. As its prolific vegetation decomposes, it feeds phytoplankton, which in turn feed algae, small fish, crustaceans, larger fish, and shorebirds. Larval shrimp, mussels, periwinkle snails, and blue and fiddler crabs proliferate in marshes where it grows. East Coast oysters anchor to the edges of spartina meadows, eventually forming dense colonies that help stabilize the shorelines.[4] Commercially important fish like menhaden and mullet use spartina marshes for their nurseries. In Texas, the Audubon Society considers the spartina-based marshlands at Bolivar Flats "the crown jewel of shorebird habitat," and populations of the endangered clapper rail feed and nest in dense spartina stands along the eastern seaboard.

Over the past two decades, as spartina started spreading rapidly in West Coast estuaries, it began dying in East Coast salt marshes. In 2000, Louisiana governor Mike Foster declared a state of emergency after 17,000 acres of spartina marshes died off within the year, with another 90,000 acres at risk. Wetland and coastal scientists on the East Coast fear the spread of so-called "brown marsh" syndrome, in which the ecosystems that once nurtured dense stands of spartina convert to open mudflats that are "simply waiting to be washed away."[5] Researchers are unsure exactly what is causing this spartina die-off, but most research points to waterlogging due to increased tidal submersion, long-term drought stress, increased soil salinity, and lack of adequate refreshment of sediments.[6] Restoration ecologists from Mississippi to New York propagate and plant spartina to save treasured habitat and vital ecological resources.

So on the East Coast, ecologists prize spartina for the diverse estuarine habitat it creates, while on the West Coast, spartina is blamed for destroying the same type of habitat. While scientists on one side of the continent scramble to figure out how to maintain populations of spartina, their counterparts on the other side of the continent work to decimate them. Is spartina an invasive species or one to be cherished, celebrated, and planted? Apparently it depends on whom you ask and what side of the country you're on.

The Reductionist Approach

Schoolchildren learn that the scientific method is based on objective reasoning and the pursuit of truth free from bias, including that of the researcher. But in the field of invasion ecology, there's a lack of consistent definitions and objective observation in the existing body of research. Although adjectives like "invasive," "noxious," and "aggressive" are used to describe plants and animals that are thriving in places where people are unaccustomed to them, ecologists disagree about the meaning and application of these terms. To date, there is no unambiguous, scientifically defensible definition of what constitutes an "invasive species."[7]

The most widely accepted definition of an invasive species is one "whose introduction causes economic or environmental harm or harm to human health."[8] But, again, this definition, which happens to be from the USDA, is more subjective than objective. Whose economic health? Does it refer to enterprising small farmers developing cottage industries based on available resources or, more likely, to large multinational agribusinesses that need to eradicate "weeds" in order to maintain their bottom line? And how do you measure environmental harm or harm to human health, especially relative to unknown, but likely damaging, effects of massive herbicide-based eradication campaigns?

On the other hand, in a 1993 report, the Office of Technology Assessment, whose mission was to provide Congress with objective analyses of technical information, acknowledged that "Some NIS [non-indigenous species] are clearly beneficial. Non-indigenous crops and livestock – like soybeans (*Glycine max*), wheat (*Triticum* spp.), and cattle (*Bos taurus*) – form the foundation of U.S. agriculture."[9] Although these three species are economically valuable (to some people), that's not necessarily true as far as their environmental or human health impacts. Farmers plant nearly 200 million acres of land annually to grow wheat and soy in the United States every year, much of this using synthetic fertilizers and pesticides. These 200 million acres were once prairie, wetland, steppe, and riparian gallery forest – considerably more biologically diverse than the monoculture fields they've been replaced with.

It's difficult, maybe impossible, to assign an objective value to living systems, which consist of complex relationships and constant change as organisms transform, and are transformed by, their ecosystems through respiration, reproduction, death, and decomposition. The intensity and scale of change ebbs and flows depending on factors such as nutrient availability, seasons, and the presence of disease or predation. As a result, any assessment of environmental "harm" requires a careful analysis of subjectivity: When a lion hunts and kills a gazelle, the dead gazelle experiences harm, but the lion, the gazelle's herd, and the grassland where they graze may benefit.

In fact, most primary research into invasion ecology is sound, objective science. It's the interpretation of the results that often lacks objectivity. Researchers rarely use the word "harm" to describe their findings; instead they apply terms like "change," "shift," "alter," and "transform" to describe the effects of invasive species. But if you read the title of a study about spartina's characteristics in West Coast estuaries such as "Changes in Community Structure and Ecosystem Function Following *Spartina alterniflora* Invasion of Pacific Estuaries," or "Invasive Cordgrass Modifies Wetland Trophic Function,"[10] although the research is presented objectively by correlating invasive species with changing ecological conditions, there's an assumption that the changes are negative and that they are caused by the invasive species in isolation from other factors. This assumption is subjective.

These studies present findings that show how spartina invasion leads to changes in the types of organisms that live in estuarine soils, from algae that proliferate in open, sun-soaked mudflats to detritivores that feed on and help decompose spartina's ample biomass.[11] The research also shows that spartina's thick root masses support the proliferation of burrowing animals like worms and larvae instead of suspension-feeding organisms like clams and snails that live near the surface of unvegetated mudflats.[12] These results suggest that the proliferation of spartina in West Coast estuaries is leaving shorebirds bereft of food. But other research has shown that shorebirds employ a variety of feeding strategies and exhibit a range of food preferences. They may feed on mudflat-dwelling snails or root-bound burrowing worms depending on what is available.[13] This diversity in feeding strategies makes sense for migratory birds that come upon different habitats and food resources depending on where and when they land. Although this is well supported by the available literature on shorebird biology, it's often overlooked in the context of spartina.

In fact, the literature of invasion ecology lacks research that definitively demonstrates that spartina harms shorebirds by reducing their food supply. Even the Washington Department of Fish and Wildlife – one of the partners in the 2003 spartina eradication campaign – acknowledges that a relationship between spartina invasion and declines in shorebird populations has not been established, yet they recommended its widespread eradication.[14] Objectivity in design and interpretation of research is paramount in the application of the scientific method, and it appears as though the field of invasion ecology is in need of an infusion of unbiased and comprehensive research. As Macalester College Professor of Ecology Mark Davis relates:

> The problem of preliminary conclusions and tentative statements being transformed into invasion gospel is a challenge the field of invasion biology needs to work hard at preventing. If invasion biology is to be a highly regarded scientific discipline, its primary assertions and conclusions need to be based on comprehensive and thoroughly vetted data sets. When conclusions are preliminary or based on data from a particular region, ecosystem or type of organism, they need to be presented as such, and those that cite these preliminary or limited conclusions need to portray them accurately as such.[15]

Unfortunately, studies that are limited in scope abound in the field of invasion ecology, and results are often not presented in ecological context. For example, in a prominent study published by Dr. Kim Patten, one of the leading advocates of spartina's removal from Willapa Bay, Dr. Patten and his team observed four sites to determine how shorebirds responded to different treatment methods. One site was a bare mudflat where spartina had never grown; another site had been covered with spartina, but was rototilled prior to the study, leaving mostly dead stubble; one was a spartina meadow that had been sprayed with herbicide; and the last was an intact spartina meadow. The researchers counted birds every week for ten minutes at each site over the course of just over a year, encompassing major migration periods for shorebirds. Shorebirds frequented both the bare mudflat and the tilled site, and researchers counted the fewest in the untreated spartina meadow.[16] Although shorebirds occurred at markedly lower (41–98 percent) densities on the sprayed site compared to the tilled site, the authors concluded that shorebird use of the ecosystem would "dramatically" increase after mechanical or chemical removal of spartina.[17]

Another study by a graduate student at Evergreen State College in Washington found very different results. Jared Parks set out to answer a question similar to Patten's: Do shorebirds use spartina, and if so, in what numbers? Parks and a research assistant counted shorebirds over the course of several hours at five different sites, which were selected based on their location throughout the bay, their accessibility, and their total spartina cover (20–80 percent). Over the course of four weekends, he counted 10,825 shorebirds, including dunlins, plovers, and teal feeding and roosting in spartina at five sites in the Willapa Bay.[18] In three instances, Parks found large, mixed groups of shorebirds preferentially feeding and roosting in spartina meadows although mudflats were exposed nearby. Parks' observations corroborate the findings of two other studies conducted in Washington in the early 1980s, which found that spartina growing in mudflats does change shorebird feeding habits, but not in a way that's clearly indicative of harm to the birds or the ecosystem.

Parks also noticed that shorebirds used spartina meadows 6.5 times more frequently during the falling tide than during the rising tide. He found the birds actively used the edge of the grassy meadows for feeding and flew into the dense stands of spartina when bald eagles and a peregrine falcon flew near. Although Dr. Patten's study offers no insight about the impact of rising or falling tides or the ways that shorebirds may use the spartina for hiding or roosting, his findings are still widely cited as justification for ongoing spartina eradication efforts.

Other research shows that spartina serves important habitat functions, even for endangered species. Although spartina is blamed for displacing shorebird habitat, recent research has found that over the course of three years, California clapper rail survival was 73 percent greater in spartina-dominated marshes than in those covered by native vegetation due to increased cover from predators during "tidal extremes" that inundate the lower-growing native vegetation.[19] In light of these findings, in 2011, the US Fish and Wildlife Service (USFWS) prohibited spartina eradication efforts (including herbicide application and mowing) at twenty-six locations in the San Francisco Bay due to concerns around the potential impacts on the clapper rail populations.[20] In 2013, the USFWS lifted the ban on all but eleven sites; for the rest, herbicide and mowing treatments are now carefully timed around clapper rail nesting season to ensure that the fledglings have left before the destruction of their nesting habitat commences.

Although spartina is blamed for threatening the habitat and food supply of shorebirds, it's the oyster production in the Willapa Bay that probably deserves more scrutiny. Oyster farmers routinely apply the insecticide carbaryl to kill native invertebrates species like ghost shrimp and mud shrimp, which burrow in and around the seed oysters, covering them with sediment and suffocating them. The EPA classifies carbaryl as highly toxic to salmonids, crabs, stoneflies, and water fleas. One study showed that carbaryl is highly toxic to mudflat-dwelling clams and snails, but that burrowing invertebrates like polychaete worms – the same kind found in dense spartina root masses – are not affected by its application.[21] And though the EPA classifies carbaryl as "practically nontoxic" to birds, it mentions that chronic exposure to carbaryl results in "adverse reproductive effects, including decreased number of eggs produced, increased number of broken eggs, and decreased fertility."[22] As concerns about carbaryl's toxicity increase, oyster producers are turning to imidacloprid to kill burrowing shrimp. Imidacloprid is a systemic insecticide that causes eggshell thinning and is classified by the EPA as "highly toxic" to many species of birds.[23] Mallard ducks, which are used in tests to determine toxicity to shorebirds and other waterbirds, showed symptoms of toxic effects, including lack of coordination and immobility even at very low doses.[24]

In Willapa Bay, the story of spartina and its ecological effects is more complex than most research acknowledges. There is little research available about how carbaryl and imidacloprid might affect shorebirds and their food sources in Willapa Bay. And yet, a rapidly spreading marsh grass is considered more of a threat to shorebirds – despite a lack of definitive evidence supporting this claim – than the widespread use of pesticides known to be toxic to the same class of organisms. Invasion ecology has incredible potential for making sense out of the dynamics of changing ecosystems, but it requires contextualization of its findings, especially if the research is used as a basis for habitat enhancement and restoration activities.

Evaluating Harm

Despite a lack of evidence for the necessity of spartina eradication, in 1995, a multiagency task force formed to develop a plan to remove spartina from Washington's estuaries. The effort is headed up by the Washington State Department of Agriculture (WSDA) and includes representatives from the Departments of Natural Resources and Fish and Wildlife, The Nature Conservancy, county and tribal governments, and researchers from the University of Washington and Washington State University. Between 1995 and 2003, these groups spent more than $10 million on spartina removal efforts including hand pulling, mowing, crushing, digging, tilling, disking, release of biological control agents, and spraying of the glyphosate-based herbicide Rodeo. Rodeo is approved for use in aquatic ecosystems because of its surfactant profile, but it was not as effective as the program's organizers had hoped it would be.

Largely due to pressure from ecologists and others working on the spartina eradication program, the imazapyr-based herbicide Habitat was approved for use in Washington in 2004.[25] The treatment protocol for most spartina meadows called for mixing Habitat with Rodeo at an application rate of 60 pounds per acre when applied via amphibious vehicle.[26] Five thousand acres of spartina were treated according to this protocol in 2004 and 6,000 more in 2005. Since then, state, federal, county, and tribal employees have eradicated the remaining patches of spartina along Washington's 3,000 linear miles of crenellated coastline and island edges. The WSDA reported in 2013 that populations of the plant have been reduced by 99 percent from their peak in 2003.[27] By 2009, the cost of the spartina control project in Washington – mostly funded by taxpayers – reached $30 million.[28] The WSDA and other agencies have spent close to $1 million per year each year since then.[29]

According to the EPA, "thousands of species of birds, mammals, fish, and other wildlife depend on estuarine habitats as places to live, feed, and reproduce. And many marine organisms, including most commercially important species of fish, depend on estuaries at some point during their development." Imazapyr and glyphosate have not been tested on all species of wildlife that use estuaries, and the mixture of Habitat and Rodeo has never been tested for toxicity or other synergistic effects. As we know, most toxicity tests are taken after only ninety-six hours of exposure, and their toxicity to salmon and rainbow trout varies depending upon the surfactant used. Imazapyr and glyphosate are persistent in soils beyond their application. If and how they are broken down or accumulated in aquatic ecosystems is unknown.

Some invasion ecologists argue that invasive species pose unknown risks. One study concluded that "not only is it difficult, if not impossible, to imagine all risks, but there is also a limit to how quantifiable certain risks are and difficulty in specifying the time frame over which impacts might be expected."[30] We must apply these same qualifications to the herbicides used to manage invasive species. Although both Habitat and Rodeo have undergone the requisite testing to determine their expected level of toxicity to aquatic organisms, and have been found to be minimally toxic, this does not mean they are safe. As we saw in chapter 1, safety testing and toxicity testing are two different approaches to determining how a substance interacts with organisms and the environment. Safety testing relies on the precautionary principle and seeks to ensure that the compound does not cause harm, whereas toxicity testing determines an acceptable level of harm. The Nature Conservancy's "Guidelines for Herbicide Use" states that "one should be confident that the proposed herbicide will do more conservation good than harm,"[31] but how is "harm" being measured?

And if the "good" is the restoration of native plant and animal communities, there are few long-term studies available that support how herbicide-based eradication achieves this goal. The few that have been conducted show that the opposite can be true. Research from Montana showed that a single aerial application of the herbicide picloram targeting the invasive forb leafy spurge decreased populations of native forbs and increased the proliferation of leafy spurge for the duration of the study – sixteen years after spraying.[32] Even researchers who promote the eradication of spartina to enhance shorebird populations admit that once spartina meadows "have transitioned to stable salt marshes, there is little likelihood that they would ever become functional mudflats again," but they still recommended broadscale herbicide application to attempt to force the transition.[33]

There is no guarantee that the herbicide-based treatment will control spartina over the long term, because herbicide-based eradication is not based on long-term solutions. Though the aggressive treatment of spartina reduced its presence in Willapa Bay from 9,000 acres in 2003 to 2 net acres in 2011, the spartina invasion may be the first of many to come – a mudflat will not necessarily stay a mudflat forever, and increased sedimentation, sea-level rise, and ocean acidification will continue to act on this estuarine environment, transforming it in unknown ways. Will the mudflats of Willapa Bay require management using broadscale herbicide application in perpetuity? What if spartina, or something "worse," comes back to fill the void?

In the time since preventing "harm" was used as the pretext for eradicating invasive spartina, other groups have starting using the same subjective justification to suit their own eradication campaigns (and, possibly, commercial interests). In 2011, following intense lobbying from the state's shellfish producers, the Washington Department of Fish and Wildlife removed eelgrass from its list of priority habitats and species. The following year, the Washington State Noxious Weed Control Board classified it as a Class C noxious weed in clam beds. After its status was changed, clam farmers in Willapa Bay petitioned the state of Washington for permission to manage it using the herbicide imazamox. The Department of Ecology granted this request, although now each producer must apply for a pollutant discharge permit before using the herbicide because of its known toxicity. In 2013, the state board lifted the "clam bed only" restriction, making Japanese eelgrass a Class C noxious weed wherever it is found in the state. Though clam producers still have to apply for a pollutant discharge permit to apply imazamox on the eelgrass, its listing as a Class C noxious weed means that a county may decide to enforce control measures – including the use of herbicide – as it sees fit. This despite clear evidence that Japanese eelgrass is a highly ecologically valuable species in Washington.[34] Meanwhile Willapa Bay, known as one of the cleanest estuaries in the world, is quickly becoming a place where chemically based marine agriculture and invasive species management is affecting the pristine character of a unique ecosystem.

The Whole Systems Approach

Amid countless reports that detail how spartina is changing the ecosystems where it is found, there is little discussion regarding *why* the invasion occurred in the first place. Spartina is known to have existed in Willapa Bay since at least the 1890s, but even the most knowledgeable researchers have no idea why the grass started to spread over thousands of acres between 1995 and 2005.[35] Thousands of studies delve into the physical characteristics of a particular invasive species, demonstrating knowledge of minute details like leaf shape and the number of dissolved oxygen molecules on their root hairs, but when it comes to how these relate to larger ecosystem dynamics – the ecological drivers of an invasion to begin with – the science of invasion ecology is strangely silent.

The organism-centered analysis so prevalent in the literature of invasion ecology leads to organism-centered management, revolving around the eradication of specific species in an attempt to return ecosystems to their former diversity, abundance, and resilience. Although there is little research suggesting this works, eradicating invasive species is considered normal and necessary within the framework of contemporary ecosystem restoration. Although many restoration ecologists know that there are larger ecological processes at work in the context of any invasion, invasion ecology as a discipline remains rooted in the idea that invasive species are the drivers of, rather than passengers on, the seemingly runaway train of ecosystem change. A more holistic approach would look at the dynamics of the invaded ecosystems as interdependent with the invading organism rather than focusing on details like whether a particular plant contains more or less vegetative biomass than another plant that lived there before.

A lot of the research on invasive species is useful and relevant. It's the context and the application that needs to change. Research should not be used to isolate and demonize certain species. It should be used to deepen awareness of the ways in which a given species is behaving in its ecosystem. The word system comes from the Greek word *synisthetai*, meaning to place together. A systems-based approach to invasion ecology places together as many relevant observations and in-depth scientific analyses as possible in order to tell a complete story of the invasion in order to offer the most holistic, long-lasting restoration outcomes. Biologist Ludwig von Bertalanffy, who developed general systems theory, wrote:

> It is necessary to study not only parts and processes in isolation, but also to solve the decisive problems found in the organization and

> order unifying them, resulting from dynamic interaction of parts, and making the behavior of the parts different when studied in isolation or within the whole.[36]

For every invasion, there is an ecological story to be told, a web of relationships waiting to be unraveled, involving contributions from multiple academic disciplines and reaching back through time. Building a whole systems–based understanding of species invasions means taking a close look at every strand in this web and piecing together evidence from historic changes in land use and their ecological results, as well as the economic, political, and social factors driving these decisions. Seeing invasive species as related to the ecosystems where they are found would represent a revolution in the science of invasion ecology and the restoration protocols it underpins. In his study of scientific revolutions from Galileo to Copernicus to Einstein, philosopher Thomas Kuhn notes that there are often two different worldviews operating at the same time in the beginning of a scientific revolution. Kuhn points out that

> practicing in different worlds, the two groups of scientists see different things when they look from the same point in the same direction . . . both are looking at the world, and what they look at has not changed. But in some areas they see different things, and they see them in different relations one to the other. That is why a law that cannot even be demonstrated to one group of scientists may occasionally seem intuitively obvious to another.[37]

The paradigm of reductionist science sees invasive species as the problem, while a systems science approach begins with an understanding of the environmental, social, and economic factors underlying the invasion and seeks to address the challenges they present from this basis. A whole systems understanding of invasive species considers them *from patterns to details*, a permaculture principle taught by David Holmgren, one of the originators of the concept. Permaculture is the study of patterns, relationships, and interconnections – the ecological application of complex systems. Although modern science contributes a vast amount of information about specific subjects, the ways in which the information is shared and interpreted tend to segregate the results into the findings of a particular discipline or department, rather than integrate the results into a complete understanding. A holistic approach to invasion ecology would integrate research from ecology, zoology, chemistry, economics, history, political science, and other fields to establish a context for an invasive organism, and move from an organism-centered to a systems-centered understanding of its spread.

In the case of spartina in Willapa Bay, piecing together findings from available research on the marsh grass shows how patterns of ecological change surrounding the invasion of spartina intersect with the details of its particular characteristics. *Spartina alterniflora* is more salt-tolerant than the spartina species that are native to the West Coast. A systems-based approach could inspire greater research into whether salinity levels in estuaries where spartina thrives are changing. Have salt levels increased in the Willapa Bay estuary over the last thirty years? How and why? Spartina is known as an "ecological engineer" that traps massive amounts of sediment. Have sediment inflows into Willapa Bay changed during the course of the invasion? Has cultivation of oysters and other shellfish transformed native marsh floral and faunal communities in the bay?

A cursory glance at some of the ecological transformations coinciding with the invasion of spartina demonstrates the importance of understanding the big picture for determining if, how, and when to manage invasive species. At least 85 percent of historic wetland and estuarine ecosystems formerly home to millions of shorebirds have been lost to urban and agricultural development along the West Coast since the 1800s.[38] Climate change–driven increases in sea levels are likely to significantly reduce the remaining intertidal habitat where shorebirds dwell.[39] In fact, we might find that spartina is beneficial to shorebird habitat since these spartina-based estuaries might form the basis of intertidal habitat as sea levels continue to rise.

Modern estuaries are repositories for agricultural and industrial pollution that inevitably flows downstream and ends up in the sediment profile. Indeed, these pollutants represent the greatest threat to existing populations of shorebirds – a greater threat than a prolific grass.[40] Infill of sediment, trash, and other fill into the San Francisco Bay has reduced its size by at least 250 square miles since 1850.[41] In Willapa Bay, the seven major rivers that drain into the estuary are prime tracts of industrial forestland where clear-cutting is the norm and flooding is a regular event. The Columbia River once contributed both freshwater and sediment flows to the bay before dams changed these dynamics.[42] The wetlands and estuaries where spartina now thrives have endured intensive industrial development over the past few centuries and today are vestiges of their former diversity, abundance, and functionality. Some researchers believe that the proliferation of spartina in estuaries worldwide is a result of these types of human-influenced changes aquatic and terrestrial ecological dynamics.[43]

We need to adopt a long-term perspective on invasions and contextualize their apparent threats and benefits in terms of larger processes whose effects are widely known, easily proven, and readily apparent. There is nothing subjective about the progressive increase of atmospheric carbon dioxide, destruction of wetlands and other critical habitats, incrementally rising seas, and the environmental fate of industrial pollutants. Focusing on eradicating spartina and other invasive species masks the difficulty of addressing the multiple and serious challenges posed by climate change, habitat degradation, and widespread pollution engendered by our modern civilization. A holistic invasive species management strategy addresses these complex processes and so has the potential to achieve truly restored ecosystems, rather than land prone to perpetual reinvasion and the need to be "controlled" by herbicide or other eradication measures.

Acknowledging our role in the degradation of the world upon which we depend is not easy. Ecologist Aldo Leopold wrote:

> One of the penalties of an ecological education is that one lives alone in a world of wounds. Much of the damage inflicted on land is quite invisible to laymen. An ecologist must either harden his shell and make believe that the consequences of science are none of his business, or he must be the doctor who sees the marks of death in a community that believes itself well and does not want to be told otherwise.[44]

In order to make the best decisions surrounding invasive species management, we must all undertake such a courageous ecological education and, in so doing, become familiar with the ways ecological systems work, the ways they change, and how our constant and inevitable interaction with them shapes their nature.

It may seem more straightforward to mow, burn, bulldoze, or spray herbicide on a plant like spartina until it disappears (for a time) than to address upstream logging or agricultural practices that may contribute increased sediment loads into wetlands or to work to mitigate increasing estuarine salinity by tackling hydrological transformations related to the interplay of dams, diversions, and wells. But the "harm" that invasive species cause must be placed in the context of contemporary large-scale ecological transformations, many of which humans are responsible for. If we really want to save shorebirds, oysters, and the wetlands and estuaries they depend upon, then we must shift our focus from addressing the visible symptoms of ecological transformation to the underlying processes that are contributing to their demise.

CHAPTER 3

Thinking Like an Ecosystem

> *"The more we study the major problems of our time, the more we come to realize that they cannot be understood in isolation. They are systemic problems, which means that they are interconnected and interdependent."*
>
> – Fritjof Capra

Life is organized in systems, their component elements and processes related to and nested within each other on infinite scales. From the universe, to the galaxy, to the solar system containing Earth and other planets held in thrall to the gravitational force of the sun, every element is related to another. Earth itself is made up of terrestrial, aquatic, and atmospheric systems, which are made up of living and nonliving elements that constantly cycle through processes of decomposition, erosion, accretion, and dissolution.

Each organism on the planet is a system with its own internal metabolic processes. The nerves, organs, muscles, blood, and lymph that make up the bodies of scallops, macaques, and rhinoceroses represent systems within systems; each is imbued with unique metabolic processes that ensure their continuance. These systems are made up of cells, which are also systems containing a nucleus, mitochondria, ribosomes, vacuoles, and other structures that are *also* systems – each with unique functions in the process of providing for life.

These tiny organelles are composed of even smaller systems, including the spiraling double helix strands of DNA and RNA that contain the genetic basis of all life. These complex and unique structures respond to changes in the cellular environment and serve as the basis for adaptation and survival. These systems are composed of tinier systems – the elements of carbon, hydrogen, phosphorous, nitrogen, and oxygen that form the basic molecular structure and the materials that bind DNA and RNA together. Each of these elements is composed of atoms, which contain protons, neutrons, and electrons, which, in relationship to each other, form the basis of all known matter in the universe. Thus the behavior of the smallest element has effects upon the structure of the largest conceivable whole. The interactions across scales are infinite and impossible to quantify, yet their intrinsic relationship to each other is definite.

In order to comprehend the behaviors of a large, complex system, it is necessary to analyze the smaller systems that make up its essential nature. It is especially important to realize that although the component pieces may be physically smaller, they are no less critical to the smooth functioning of the greater system. This concept is expressed through the idea of "nested hierarchies," in which layers of functional groups, each with their own internal autonomy, function together to ensure the optimal performance of the whole system. Hierarchy in this sense does not refer to control over and can instead be thought of as a means of delegating control and responsibility over actions and decisions appropriately scaled to the managing system.

For example, the human brain is often considered the most critical organ in the human body, the director in charge of the processes that control life. However, once removed from the body – disconnected from the heart that pumps blood, the liver that filters toxins, the protective casing of the skull – the brain is merely a lump of lifeless matter. It is not that the brain, heart, liver, or skull is any more or less important than the other. Each has its function within the context of a living body. They are instead distinct, functional parts of a whole, reliant on their component parts, and the whole body system, which is in turn reliant on the functioning of even larger systems like those that provide food, water, and air.

The Nature of Systems

Systems are found everywhere, but not everything is a system. In order to understand the differences, systems scientist Donella Meadows describes systems as having the following qualities:

- Integrity or Wholeness
- Adaptive
- Resilient

- Evolutionary
- Goal-Seeking
- Self-Preserving
- Self-Organizing

Every system has all of these characteristics. Consider a pile of leaves under an oak tree on an autumn day. Unto itself, the pile is not a system – yet. The leaves may scatter, and the pile does not exemplify traits of evolution and goal seeking. It does not have an internal mechanism governing its pileness. However, the tree that dropped its leaves does meet the criteria of a system. It evolves in response to changes in climate and season and shows traits of self-preservation by healing wounds and repelling insect damage. The soil in which the tree grows is also a system. The organisms and elements in the soil are resilient and adapt to changing conditions. If the leaves are left to decompose under the tree, they become a part of this soil system. If they are raked up and burned, they become a part of the system that circulates carbon throughout the biosphere.

When it comes to nonhuman systems, "goal-seeking" is different from how we might initially think of it. That a system is driven toward a goal or has a purpose is not necessarily a matter of conscious intent. The "goal" of the circulatory system of a tree is to carry sap throughout the living tissue, transporting nutrients and preventing infestations of disease-causing organisms. The tree does not necessarily set this goal in the same way a person would set a goal of jogging one mile every day, but the circulatory system will accomplish these tasks, barring disease or a major disturbance. Goals in a system are principles around which activities and processes are organized.

Systems are self-organizing and self-preserving and maintain these traits through feedback processes that encourage and enable systems to remain whole and functional. In the human body, this is known as homeostasis, whereby the body system stays alive by maintaining a dynamic equilibrium through the regulation of both negative and positive feedback processes. These feedback processes are finely calibrated to ensure the continued existence of the living human body; if they are excessively amplified or inhibited, it can result in death. But even death represents the recycling of matter into other systems.

The feedback processes that characterize systems operate through constant interaction and interplay among various elements and processes. Positive feedback is an additive, reinforcing process. In the human body, blood clotting is an example of a self-reinforcing positive feedback loop. When you are wounded, platelets flow to the site of the wound, sending chemical signals that attract more and more platelets until they reach a critical mass that can staunch the flow of blood.

Negative feedback loops slow down or stop processes that are detrimental to the functioning of the system. In natural systems, they're more common than positive feedback loops since there are many external influences that may need slowing down or stoppage to avoid harm or death. For example, the body's production of insulin following a blood sugar spike is a negative feedback response. When someone eats sugar, it signals the pancreas to release insulin, which brings blood sugar down to stable levels by driving sugar into certain cells for storage. If blood sugar drops, the body releases this stored energy in another negative feedback loop that ensures the body can perform basic functions until the next meal arrives.

The planet as a system is also prone to feedback loops that reinforce or inhibit different processes, many of which influence global climate patterns over time. For example, bright white frozen ice caps reflect solar energy back into the atmosphere. But when atmospheric temperatures rise and ice caps melt, the newly exposed dark ocean water traps radiant heat from the sun – a positive feedback loop that leads to further warming of the ocean's water. As warming continues, negative feedback processes – such as the rising ocean temperatures leading to increased evaporation and more cloud cover, which increases the amount of sunlight reflected back into space – could balance the positive feedback process represented by increased solar energy (and warmth) stored in the planet's oceans. But as temperatures continue to rise, melting ice and permafrost release methane – a greenhouse gas twenty times more potent than carbon dioxide – further accelerating climate change in a strong positive feedback loop. Climate scientists predict that if the release of methane reaches a certain point, it will destabilize the climate, moving us toward an uncertain future of accelerated warming, with unknown potential negative feedbacks. Throughout Earth's history, increases in temperature and carbon dioxide have triggered ice ages – negative feedback processes that counteract the warming.

Independent scientist James Lovelock has observed that conditions on the planet have remained fairly constant and hospitable throughout the long history of life on Earth – despite these complex and incredibly dynamic feedback processes (or maybe because of them). For example, the oceans have maintained a relatively stable salinity profile despite constant inputs of salts from terrestrial erosion and weathering. Earth's atmospheric composition has also remained relatively consistent, especially over the last 2.5 billion years. And although the amount of energy emanating from the sun has increased 25 percent since the formation of the solar system, surface temperatures on the planet have not followed suit – if they had, life would have been obliterated long ago.[1]

Lovelock worked with microbiologist Lynn Margulis to develop these observations into the Gaia theory, which explains how life on Earth interacts with Earth's inorganic matter to create a self-regulating, homeostatic system. Lovelock and Margulis emphasized that the Gaia theory does not argue that the planet is a sentient being, but argues instead that "the sum of planetary life, Gaia, displays a physiology that we recognize as environmental regulation. Gaia itself is not an organism . . . it is an emergent property of interaction among organisms, the spherical planet on which they reside, and an energy source the sun."[2] The Gaia theory defines the planet as a system – an integrated, evolutionary, goal-seeking, adaptive, self-organizing, and self-preserving whole.

Invasible Ecosystems

Ecosystems are systems – even when they're invaded. A species invasion is an opportunity to investigate whether and how a transformation in other ecological relationships and feedback processes might be facilitating the invasion. In order to understand what makes an ecosystem even prone to invasion, we have to take a close look at what is happening in the places where invasive species thrive. Ecologists Marcel Rejmánek, Petr Pysek, and David Richardson found that invaded ecosystems around the world exhibit one or more of the following traits: a significant disturbance in their history, resident vegetation with slow recovery rates, a significant change in nutrient accumulation or dispersal, and/or a fragmentation of successionally advanced communities.[3] In other words, invasive species thrive where ecological conditions have changed.

Even small or short-term changes in resource availability can make an ecosystem more vulnerable to invasion. One study found that limiting water in an experimental plot of the native forb black-eyed Susan for one month decreased its seed production and dispersal the following year, making its ecosystem more prone to invasion by nonnative grasses.[4] Although this was not a longitudinal study, it demonstrated that even minor fluctuations in resource availability can have significant and lasting effects.

On the other hand, ecosystems, like all systems, are dynamic. They're *characterized* by constant change. Ecosystems, like any system, can sustain and even thrive with some level of perturbation, but at a certain threshold, they are prone to complete transformation. Donella Meadows explains, "A system generally goes on being itself, changing only slowly if at all, even with complete substitutions of its elements – as long as its interconnections and purposes remain intact. If the interconnections change, the system may be greatly altered. It may even become unrecognizable."[5] If so many elements and their interconnections, which give stability and resilience to the system, are changed or lost, all bets are off as far as what form the system takes next. An unrecognizable system isn't necessarily dysfunctional – it's just different from what it was.

So invaded ecosystems are systems that may be greatly altered or even on the brink of becoming unrecognizable. In invaded ecosystems, the relationships that once supported a particular array of plants and animals, or a certain manner in which nutrients and water cycled – those particular ecological attributes that have come to be known as native or "normal" (or recognizable) – are gone, replaced by others that may seem less than ideal. We mourn the loss of biodiversity or ecological function – and we may feel deeply unsettled by a sense of unfamiliarity with a landscape we thought we knew. And yet, those invasive species may be remarkably well suited to the conditions where they thrive and may be perfect complements to changed and changing conditions. The Colorado River, for example, has been greatly altered and is perhaps on the brink of being unrecognizable, at least if you observe it in terms of the organisms – such as salt cedar – that thrive there today.

Salt Cedar on the Colorado River

The Colorado River begins as snowflakes and raindrops that fall on the Rocky Mountains, where moisture accumulates as it is driven north by winds from the Gulf of Mexico. The river winds its way from these rocky crags through sandstone canyons and desert plains to its mouth in the Sea of Cortez. The first year that water from the Colorado River was diverted from the watershed was 1901, flowing through a canal into California's Imperial Valley to irrigate the state's fledgling agricultural industry. In the 1930s, the All-American Canal, the longest irrigation canal in the world, was constructed for the same purpose – to provide water for 630,000 acres of produce grown in a region with little natural irrigation. The All-American Canal has moved more than 26,000 cubic feet per second (cfs) – a flow equivalent to the flood stage of a major river – of water through the sandy desert of Arizona and southern California every day for more than eighty years.

In 1935, a six-company joint venture completed construction of the Hoover Dam along the Colorado River at the border of New Mexico and Arizona. Construction on the Glen Canyon, Painted Rock, Flaming Gorge, and other dams began in 1956. Today, the 1,450-mile length of the Colorado River is obstructed by fifteen dams. Hundreds more line its tributaries, providing electricity and irrigation and municipal water for thirty-five million people.

Prior to construction of the dams and canals, the Colorado River had one of the most wildly fluctuating river flows in the world, from seasonal highs of over 100,000 cfs during spring snowmelt – equal to the flow coursing over Niagara Falls – down to a few thousand cfs in late summer. The remarkable erosive force of the river is strikingly visible in the Grand Canyon, where the river has worn its way through brightly colored sandstone bedrock more than a mile deep over the course of several million years.

The bands of colorful rock through which the Colorado River flows are marine sediments dating as far back as five hundred million years ago, which give the river a natural salinity. However, industry, especially agriculture, increases that salinity. When farmers overirrigate with more water than a plant can use, the excess salty water evaporates in the hot, dry climate, leaving salt deposits on the soil surface where they build up over time. Inefficient irrigation also pushes excess salty water into the water table, driving dissolved salts toward the root zone of the irrigated plants. And irrigation water also absorbs salt-containing fertilizer and chemical residues, further increasing water and soil salinity as runoff leaves the fields and makes its way back into waterways.

Additionally, coal-mining spoils leach salt into the river as naturally occurring salt in the soil is exposed to rain and snow when it's discarded in loose piles by the mining operations. And natural gas–drilling operations in the area also inject salty water (as well as a series of chemicals) into deep wells to pressurize and release the gas. Once the gas is extracted, the wells are abandoned, slowly leaching their contents into the watershed. And then, of course, there are the dams with their large, open expanses of water in hot, dry deserts. The US Bureau of Reclamation – the federal agency in charge of reclaiming water from "wasting" to the sea – estimates that 1.5 million acre-feet of water per year evaporate from the networks of dams and aqueducts on the river, leaving salts behind in greater concentration.[6] The EPA estimates that human activities like mining, gas exploration, and agriculture have more than doubled the salinity of the water in the Colorado River from historic levels.[7] Today, during low-flow seasons, salinity is as high as 1,000 milligrams per liter (mg/L) in the lower reaches, well beyond the limits of what most riparian vegetation can tolerate.

Although the river is naturally salty and the vegetation is adapted to saline conditions in the soil and water, salinity increased markedly after the development of these industries in the Colorado River watershed. In a 1971 study, the EPA estimated that only 47 percent of the salinity of water measured at Hoover Dam came from natural sources like erosion from marine sediments and flows from naturally salty springs. The remaining salt load was due to human activity – irrigated agriculture contributing the bulk of it.[8]

The Colorado River also carries a lot of sediment. Although today the water appears clear at least below the dams, it was once known by settlers to the area as "too muddy to drink, and too thin to plow." The river is still churning up the ancient seabed sediments; approximately forty-four million tons build up behind the wall of the Glen Canyon Dam every year and approximately eighty-four tons every minute.[9] The dams were projected to last for three hundred to seven hundred years when they were first constructed, but they are filling with sediment more rapidly than the initial engineering estimates predicted, and this rate is expected to increase.[10] Today, the Bureau of Reclamation is studying the feasibility of dredging and removing this buildup and estimates that moving the sediment from behind the dam to the delta – where it would have been deposited in the absence of the dams – will cost $2.6 billion annually in transportation costs alone.[11]

When asked in 1995 how the Bureau of Reclamation planned to address the inevitable problems associated with sediment buildup behind the dams on the Colorado River, then-commissioner Floyd Dominy remarked, "We will let people of the future worry about it."[12] The future has arrived much sooner than anticipated, and since dredging and removing the sediments is the only way the dams can remain operable over the long term – it looks to be costly.

And, of course, important industries and population centers rely on the river for their survival. Los Angeles, San Diego, Tucson, Las Vegas, and Phoenix rely on the Colorado River for municipal water supplies, as do Tijuana, Mexicali, and Ensenada. The river provides domestic water for more than thirty-five million people. Farmers irrigate 4.5 million acres of valuable agricultural land – growing crops like alfalfa and nearly all of the winter vegetables consumed in the United States – with water from the Colorado River. Dams on the river produce 12 billion kilowatt hours of electricity, enough to provide power to one million average US households for a year.

And every year, the entirety of the river's flow is spoken for before a single drop even falls in the watershed. In 1922, seven states signed a compact that apportioned 7.5 million acre-feet of Colorado River water to Wyoming, Colorado, Utah, New Mexico, and northern Arizona, as well as 7.5 million acre-feet to California, southern Arizona, and Nevada. In 1944, the United States and Mexico signed a treaty guaranteeing Mexico an additional 1.5 million acre-feet every year. In the years leading up to the Colorado River Compact, the river's flow was measured at 16.5 million acre-feet, enough water to cover 16.5 million acres – approximately equivalent to the land area of Massachusetts, Vermont, New Hampshire, and Delaware combined – at 1 foot deep. But, the flow data used as the basis for these negotiations were based on assessments that took place after a string of particularly wet years. Tree ring analyses show that the river's long-term average flow is closer to 13–14 million acre-feet per year.[13] No state wants to give up its water, and the compact has yet to be renegotiated based on the newer, more accurate data.

Today, because of overapportioned flows, most years, the Colorado River dries up a hundred miles upstream from its natural terminus in the Sea of Cortez, leaving the vast wetland delta ecosystem formerly supported by the river caked and dry. And water flow along the entire length of the river is decreasing, as well. The Scripps Institute of Oceanography predicts that reservoirs on the Colorado River may dry up by 2021 because of climate change–induced changes in snowpack.[14]

The overallocation of water flows, increasing salt loads, changing patterns of sedimentation deposition, and lack of seasonal flooding cycles has important implications for the ecology of the Colorado River watershed and other southwestern rivers. Three of the eight species of fish endemic to the Grand Canyon portion of the river are extinct, and another two are critically endangered. According to the National Park Service, these extinctions followed the completion of the Glen Canyon Dam in 1963, when the fish assemblage in the Grand Canyon stretch of the river shifted to nonnative brown and rainbow trout.[15] The dense stands of Fremont cottonwood and Goodding's willow that once shaded long stretches of sandy riverbank are gone, and the invasive salt cedar thrives throughout the watershed.

Salt cedar is known for its ability to bind fine sediments in its roots and was originally planted in the Colorado River basin at the head of reservoirs in an effort to keep the heavy silt loads from building up behind the dams. As its name implies, salt cedar is a halophyte, or salt-loving plant. As it transpires, salt cedar takes up saline water and exudes excess salt through its leaves. When the leaves drop, the salt is deposited on the surface of the soil. This accumulation of saline leaf litter is considered a deterrent to the germination of salt-intolerant native species of willow and cottonwood. However, in the few free-flowing and undammed rivers of the arid southwest – such as the San Pedro River in Arizona and the Gila River in New Mexico – that have variable seasonal flows and frequent flooding similar to the historic Colorado River, soil salinity levels are low under all vegetation types, including under salt cedar.[16] In these rivers, the regular flooding that occurs during spring snowmelt and in pulses throughout the summer from the "monsoon" rain events common in the southwest washes away the salty leaf litter.

Although the US Forest Service states that salt cedar provides little value as wildlife habitat, the *Arizona Breeding Bird Atlas* data show that forty-nine species of birds use salt cedar as a breeding habitat, including the endangered western willow flycatcher.[17] The shrubs also produce profuse nectar and early season pollen. Today salt cedar thrives along the banks of the Colorado River, and though the trees are subjected to chainsaws, bulldozers, and herbicides in attempts to control their spread, they continue to proliferate. One study modeling future climate change scenarios predicts that tamarisk populations will experience a two- to tenfold increase over the next century due to increases in habitat suitability.[18]

In 2006, Congress passed the Salt Cedar and Russian Olive Control Demonstration Act, authorizing $80 million in spending over the next four years on research and control of the rapidly spreading shrub.[19] The US Forest Service lists root grubbing, excavating, and bulldozing as viable control methods for small stands of the plant in its *Field Guide for Managing Salt Cedar* and mentions ongoing research into the effects of releasing the salt cedar leaf beetle as a biological control agent. The US Forest Service also recommends the application of the herbicides triclopyr (Garlon), imazapyr (Arsenal), and glyphosate (Rodeo) by aircraft, helicopter, tractor or ATV-mounted boom sprayers, power sprayers, backpack sprayers, and carpet rollers for large stands of salt cedar. And it recommends as the most effective strategy a combined approach, especially the "herbicide-burn-mechanical control" program, where aerial herbicide application kills 70 to 90 percent of standing salt cedar, followed by controlled burning two years later to remove dead aerial trunks and stems. If prescribed burning isn't possible, mechanical treatments such as chaining, cabling, bulldozing, or roller chopping are recommended to drop standing dead debris. Surviving salt cedar plants are removed in the fourth or fifth year after spraying with an excavator, grubber, or root plow and raking. And the Forest Service recommends further herbicide applications if the salt cedar resprouts.[20]

The Forest Service field guide recommends planting native species, especially cottonwoods and willows, after these interventions, since the stream banks are highly susceptible to reinvasion. Unfortunately, even a limitless supply of herbicides, the world's largest excavator, and all of the money in the US Treasury will not create the conditions in which cottonwoods and willows thrive, at least as long as the Colorado River maintains its current form. Willows and cottonwoods are flood-adapted species, reliant on the nourishing influx of mineral-rich sediments and periodic flushes of freshwater to remove built-up salts. Changes in seasonal flows and an increase in salinity from agriculture, dams, and mining activities have changed the character of the Colorado River as a system, and within the new system, salt cedar flourishes. Willow and cottonwood do not.

Habitat versus Niche

If we take a holistic, permacultural view of salt cedar, it appears to be playing a perfectly logical role in today's Colorado River watershed. Ecosystems are made up of niches, the functional roles that organisms play within their environment. Niches are different from habitats; they describe a relationship rather than a location. Ecologist H. T. Odum describes *habitat* as an organism's "address" and a *niche* as its "profession." A habitat provides information about where an organism lives, whereas its niche describes what it does in the place where it lives. Analyzing a species in terms of its niche offers an understanding of its presence in a habitat – why it is found in a particular place at a particular time. All organisms, including invasive species, are making use of available niches in an ecosystem. Native plants also have unique niches and are suited to particular conditions. Their failure to thrive in invaded ecosystems is not necessarily due to the invasion, but to changes in the underlying conditions that once contributed to their survival.

Analyzing an organism in terms of its niche is the first step in making an ecosystem-level analysis of a species invasion. All organisms are *doing* something in the ecosystems where they live; salt cedar and other invasive species are no exception. Although salt cedar is decried because it drops salt-laden leaf litter to the ground, consider what this action means for the ecosystem. Halophytes like salt cedar absorb salty water through their roots, accumulate salt within their tissues, and transpire freshwater through their leaves. Transpiration releases moisture into the atmosphere, and little by little, the salty water that other plants' roots come in contact with is rendered *less* salty. In an ecosystem where the water is becoming ever more saline due to evaporation and agricultural runoff, the highly salt-tolerant salt cedar is filling a niche that could not be filled by many other plants.

Salt cedar is also considered a "thirsty" tree that takes up valuable water and "wastes" it through evapotranspiration. One study concluded that the evapotranspiration rate of salt cedar in the southwestern United States uses nearly twice the amount of water every year than all of the large cities in southern California combined,[21] although the research was later discredited for using outdated methods to calculate evapotranspiration. More reliable research has shown that salt cedar has intermediate evapotranspiration rates – and that native cottonwoods and willows have the highest rates.[22]

Nevertheless, in a watershed where every drop of water is carefully metered and directed toward economically measureable outcomes like crops and kilowatts, economically "useless" vegetation like salt cedar drinking up water adds fuel to the fire of eradication campaigns. The conventional wisdom is that by clearing salt cedar and other riparian vegetation, more water could become available for other uses.

Unfortunately, salt cedar eradication efforts aimed at reclaiming water have not led to more available water.[23] Although removing vegetation might raise water tables, it also increases evaporation due to bare soil. Over time, other plants will colonize the ground and increase evapotranspiration again. In addition, raising the water table makes it more likely that surface salts will percolate into the groundwater and further inhibit the establishment of salt-sensitive native species. Though salt cedar and other riparian vegetation use water, these plants also humidify the environment through their evapotranspiration. Trees and long-lived shrubs are important denizens of arid environments because they cool soil and humidify the air. In large enough quantities, this water vapor can rise, condense, and fall back to the land as rain.

It's worth noting that evapotranspiration through plants is different from evaporation from bare soil because the water exuded from a plant's leaf surface contains more than just water. Plants also release cloud-seeding bacteria like *Pseudomonas syringi* into the atmosphere. This organism has a unique ability to encourage the formation of rain and has been found in raindrops throughout the world.[24] The shape and size of this bacteria provides a substrate for the formation of ice crystals, which eventually form together into clouds, and fall back to Earth as rain or snow. (Seeding clouds using silver nitrate is based on a similar idea – putting a substance with lots of surface area into the atmosphere provides a substrate for water vapor to accumulate on, and eventually clouds will form and a rainstorm will commence.) *Pseudomonas syringi* also encourages nucleation of ice crystals at higher temperatures than other agents, allowing for rainfall to occur even in hot, dry desert conditions if enough of the bacteria are present for water to condense on. David Sands, who originally proved the existence of such bioprecipitators, also believes that there may be a connection between bacteria like *Pseudomonas syringi* and the cyclical droughts that plague farmers in arid regions. He argues that by overgrazing or clearing vegetation from the brittle region, the bioprecipitating bacteria lack adequate leaf surface habitat from which to be swept up into the atmosphere. In one interview, he suggests that by planting species known to harbor more bacteria, farmers may be able to improve their regional water cycle and protect from drought.[25]

Salt cedar harbors more bacteria – both in terms of number and diversity – than non-salt-accumulating plants in the same environment. One study discovered sixty-eight genera, including *Pseudomonas*, on the leaf surfaces of salt cedar.[26] This study also found several species of bacteria that are not yet identified – known only, for example, as "Antarctic seawater bacterium" and "Glacial ice bacterium," after the places where they were first collected. Perhaps salt cedar shrubs, with their masses of vegetation providing habitat to hundreds of ice-nucleating bacteria species, are helping to create rain in the Colorado River watershed. If Dr. Sands is correct, and these ice-nucleating bacteria are an important part of the hydrological cycle, especially in high temperature and arid regions, then plants like salt cedar with such high numbers and diversity of microorganisms should be considered a boon to the ecosystem, where they stimulate rainfall and perhaps even the flooding so critical for the native species of plants and animals that once thrived in the watershed. And yet, despite this potentially crucial role that salt cedar fills in desert riparian ecosystems, it is still targeted for eradication.

Disturbance

As we saw earlier in the chapter, disturbance can contribute to the likelihood of invasion in an ecosystem, as changing ecological dynamics can lead to significant alterations in species composition. Many common invasive species thrive in highly disturbed ecosystems – dandelions push their way through cracks in the concrete, and tree of heaven thrives along highways and in coal mine spoils. Although some invasive species fill niches where the scale of disturbance has made it impossible for all but the hardiest organisms to survive, some disturbance is also critical to maintaining the diversity and abundance of native species. The native vegetation along the Colorado River, for example, depends on regular flooding intervals for renewal and growth. Although flooding can be intense and may lead to the death of individual trees or shrubs, on a larger scale, it makes more and different resources available. Greater biodiversity in the context of moderate disturbance is known as the intermediate disturbance hypothesis – biological diversity is highest when disturbance is neither too rare nor too frequent.[27] This is true in a variety of ecosystems, from tropical forests, to coral reefs, to streams.

Discovering the ideal scale and timing of disturbance is tricky. Too much, and the ecosystem could change completely; too little, and diversity declines. Research on the Rio Grande River has shown that simulating the flows and slow drawdown rates engendered by natural spring flooding patterns stimulated the germination of cottonwoods, even in the presence of salt cedar.[28] Another analysis of fifty-four stream sites showed that the type and scale of disturbance in the streambed was the most important indicator of plant species diversity found at those sites.[29] Riparian vegetation is very sensitive to shifts in moisture availability, flood frequency and duration, and scouring effects.[30] The lack of seasonal flooding on the Colorado River has ecological consequences – the intermediate disturbance of the spring floodwaters contributed to the former floral and faunal diversity of the riparian ecosystem. The construction of dams and other changes in hydrology on the Colorado River are also ecological disturbances, but not the sort that favor the growth of willow and cottonwood. Recall that if enough relational elements are severed, the ecosystem, like any other system, may become unrecognizable. Invasive species are evidence of this in many damaged ecosystems. When the scale and timing of disturbance changes an ecosystem to this great degree, it can create conditions that favor the growth of different species than existed in the past.

The intermediate disturbance hypothesis has important ramifications for understanding invasive species, because they don't only thrive in overly disturbed environments, but in under-disturbed ecosystems as well. Conifer trees are uniquely adapted to the periodic disturbance of low-intensity fires to stimulate germination of their seeds. In the southeastern United States, forests of longleaf pine are slowly dying after decades of fire suppression restricted successful reproduction of the species. Today, these pine forests are also host to invasive species like Japanese honeysuckle, whose vines can choke, and eventually kill, young trees. However, one study found that the reintroduction of periodic, low-intensity burning to the longleaf pine ecosystem kills Japanese honeysuckle and stimulates the germination of longleaf pine seeds present in the soil seed bank.[31] This same study found that the diversity of forbs, legumes, and grasses also increases through well-timed application of fire. The Missouri Department of Conservation considers prescribed fire one of the best approaches for eliminating Japanese honeysuckle, and although it also recommends herbicide application, it acknowledges that repeated low-intensity burning will ensure its eradication over time.

Ecologist C. S. Holling characterized the pattern of disturbance and the opportunities it creates as the adaptive cycle – how ecosystem resources, like light, water, and nutrients, are utilized in a constantly shifting interplay of interactions and feedback. The adaptive cycle explains how ecosystems behave as systems – maintaining their inherent integrity while their components transform over time. The first stage of the cycle can be observed after a major disturbance such as a volcano or clear-cut, when rapidly growing pioneer species colonize the land. This rapid-growth phase gives way to larger and longer-lived species that eventually form the dominant canopy of the ecosystem. This is known as the conservation phase of the cycle, such as in an old growth forest where there are few large trees and little growth in the understory – the ecosystem has gone through its period of rapid growth and has settled into its longer-term, more stable pattern. Toward the end of this phase, there is little room for other species to gain a foothold and survive. The available energy (sunlight, water, nutrients) is largely bound up in the well-established species. Even in the most stable system, at some point, another disturbance or release event occurs. Perhaps some trees fall in a windstorm, and sunlight reaches the forest floor for the first time in centuries. This leads to the reorganization phase, in which energy and resources formerly bound up in the conservation phase are made available to whichever species are able to take it up. The cycle begins again with the rapid-growth phase. Although ecology is cyclical, this doesn't mean that an ecosystem returns to a former state in the event of a disturbance. This point is crucial – since return to a previous state seems to be the goal of many invasive species eradication programs.

Before the Colorado River was dammed, the seasonal flooding released resources by killing older trees and shrubs, which stimulated the germination and rapid growth of cottonwood and willow on the newly deposited sediments, using the available nutrients, water, and sunlight. As long as the spring freshets and summer monsoons keep flowing, this adaptive cycle more or less continues. If these patterns shift dramatically, say through the construction of dams and highly regulated water flows, the pattern of the cycle would change. The adaptive cycle offers a means by which to anticipate that novel organisms will fill newly available niches and that these organisms will have different strategies for making a living in the transformed conditions. Enter salt cedar, whose characteristics are attuned to the available resources of a highly altered Colorado River.

Succession

Succession describes how an ecosystem changes over time as energy and resources are bound up in larger and longer-lived organisms. The first organisms to colonize an ecosystem following a disturbance are known as early successional or pioneer species. These species are often rapidly growing and short-lived, and tend to produce a lot of seeds or other propagative material. Early successional plants tend to have high rates of photosynthesis, respiration, and resource uptake and more efficient use of available sunlight, whereas late successional plants often have the opposite characteristics.[32]

Early successional species also tend to have lots of flowers, fruits, and seeds that attract other organisms, which help to bring along the next phase of succession. Pollinators like bees, wasps, and butterflies feed on nectar, and their growing populations feed birds, lizards, frogs, and small mammals like skunks and raccoons that devour their larvae-filled nests. Seed-eating birds also eat the seeds of pioneer species and then deposit phosphorous-rich manure into the soil. Over time, this fertility, as well as other organic matter from decomposing plants and the proliferation of soil biota from root exudates, contributes to the growth of the next phase of succession.

So pioneering plants do a lot to help create the conditions necessary for the next stage of succession, often characterized by shrubs and bushes. These larger, woodier plants are less likely to germinate and thrive in the barren conditions of a postdisturbance environment, but once established, these shrubs provide shelter, nesting sites, and food, as well as more shade, better soil, erosion control, and water retention. As individuals in this stage of succession become more dominant in the environment, the pioneer species are shaded out and, their ecosystem services complete, begin to die out.

At the same time, the increased shade, moisture, and nutrient availability creates perfect conditions for the seedlings of small and large trees to germinate and begin their comparatively slow growth process. This improved bird and animal habitat provides an attractive destination for wildlife, which leave droppings full of seeds encased in a nutrient-rich packet of manure. Over time, the tree seedlings grow and dominate the canopy, eventually shading out the shrubs and remaining pioneer forbs and grasses. In their deep shade, other shade-tolerant species germinate, and different birds, animals, and insects make their homes. This high-biomass, high-complexity ecosystem is known as the climax stage, and only another significant disturbance, such as a fire or windstorm, will alter its stability, starting the successional process once again.

When ecologists first described succession in the ecological literature of the 1920s, the leading hypothesis at the time was that ecosystems were the products of tightly bound symbiotic relationships of organisms that had developed in place over the course of millennia and that the changes within these systems should be logical, orderly, and predictable over time. A competing hypothesis argued that communities were not inherently structured, but rather groups of independent species specialized to survive in the ever-changing conditions to which they were best adapted. Henry Gleason, who developed this latter hypothesis, wrote:

> The vegetation of an area is merely the result of two factors, the fluctuating and fortuitous immigration of plants and an equally fluctuating and variable environment. As a result, there is no inherent reason why any two areas of the earth's surface should bear precisely the same vegetation . . . Every species of plant is a law unto itself, the distribution of which in space depends upon its individual peculiarities of migration and environmental requirements. . . . The species disappears from areas where the environment is no longer endurable. It grows in company with any other species of similar environmental requirements, irrespective of their normal associational affiliations.[33]

According to Gleason's line of reasoning, there is nothing special about a particular association of species within an ecosystem – their coevolved symbiotic relationships are a matter of chance based on the mutual advantages gained by such interactions and long periods without significant systemic change. Gleason believed that these associations were not preordained by an overarching orderly community structure and argued that the components of an ecosystem, including its organisms, cannot be predicted over time. Though Gleason's ideas were not well received by scientists at the time of his research, today the concept that ecosystems tend toward a predetermined stable climax state has largely been abandoned in favor of a more dynamic model of ecological change. Indeed, the most recent research embraces the idea that ecosystems have multiple stable states, with many potential organisms filling ever-shifting niches, just as Gleason predicted.[34]

A singularly important concept of succession is that no single species will survive for time immemorial in a given ecosystem. Every species is part of the ecosystem's successional drive, and eventually, the conditions will not be appropriate for a given organism's growth. Sun-loving pioneer plants cover ground quickly and are eventually shaded out by the shrubs and trees that germinate in the vegetative cover that they provide. When old trees fall down and allow sunlight to penetrate the forest floor, they open the door for new vegetative growth that feeds on the decomposition of their carbon-rich bodies. Nitrogen-fixing plants can survive in nitrogen-poor soils, and eventually, other species will thrive in the fertile soil that they create. Any organism present in an ecosystem is representative of its successional state, which will change again over time inevitably and inexorably.

Succession and Invasion

Most species that are considered invasive are classified as pioneer, early, or mid-successional species. The early stages of succession are more prone to invasion because there are greater resources and opportunities after a disturbance.[35] Pioneer or early successional invasive species include spotted knapweed, leafy spurge, and garlic mustard. Such species are usually annuals or short-lived perennials, consist of mostly vegetative (nonwoody) growth, and often have showy nectar-rich blossoms and prolific seed yields. Pioneering species also tend to have deep taproots, like the invasive bull thistle, soil-accumulating fibrous roots, like cheat grass, or rapidly spreading rhizomatous roots, like those of giant goldenrod. These attributes help the plants make use of newly available resources after a disturbance and also lay the groundwork for what comes next.

Early successional species thrive after a disturbance, as new resources are made available by changing conditions. Consider the preponderance of weeds in a conventionally tilled agricultural operation: The soil is disturbed at least once, possibly several times a year, in order to provide crops with light, water, and nutrients. The seeds of corn, beans, or broccoli are among many other plants that will germinate and make use of newly available resources. The noncrop plants are considered weeds because they take advantage of the resources that farmers want for the crops. But weeding (or spraying) doesn't eliminate the successional drive of the ecosystem where the farm exists, and any further disturbance makes room for weeds to sprout once again. The only way to get beyond this phase is to stop attempting to arrest succession with continued disturbance or guide the ecosystem toward the conservation phase of longer-lived perennials with carefully applied disturbance events.

Certain invasive species colonize where disturbance is not recent or obvious. Some of these include mid-successional species like kudzu, Japanese honeysuckle, English ivy, glossy buckthorn, and tree of heaven. Mid-successional species are usually slower growing perennials, including trees, shrubs, and vines. They are often woody, longer-lived, and some of them, like the honeysuckle, ivy, and kudzu, can grow in shade. They often have nectar-rich flowers as well as fruits, nuts, and seeds that are eaten by birds and mammals that thrive in young forest ecosystems. Mid-successional species may not be the first to colonize postdisturbance environments, but they are indicative of a disturbance-related successional trajectory since, like the pioneer species, they are also reorganizing resources made available by a disturbance. Mid-successional invasive species often thrive in areas where disturbance has ceased or changed in tempo from historic norms, such as the incursion of Japanese honeysuckle into longleaf pine forests of the southeastern United States where fire suppression replaced historic pulses of low-intensity burning—which facilitated the proliferation of honeysuckle and other invasive species. Salt cedar is another example; its proliferation is a consequence of the lack of intermittent flooding on the Colorado River.

There are few later successional or climax invasive species because there aren't as many available resources in later successional ecosystems. Imagine an old growth forest of two-thousand-year-old redwoods in Northern

California. This unique forest ecosystem is full to the brim with species and their interrelationships – even the canopy layer of ancient redwood trees are home to a diversity of organisms that never touch the ground. This system is at the height of the adaptive cycle's conservation phase: The ecological niches are full, and resources like light, water, and nutrients are scarcely available for new and innovative species. Imagine pouring a bag full of seeds from the world's "worst" invasive species, like spotted knapweed, Scotch broom, and kudzu, into the dense and dark redwood forest understory. They would probably decompose within months without ever seeing the light of day. If some germinated, they would find little light and few nutrients unless they were fortunate enough to partner with the extensive network of fungal mycelia that dominate nutrient exchange in such old forest ecosystems. Even these tenacious species would fail to thrive in such a successionally advanced environment – there is no room for them to do so. However, if many of these trees were cut down, within a matter of days, the dank, dark understory would be flooded with sun and exposed to wind, rain, and logging equipment. The decomposing roots below the surface would provide ample nutrients for any plant species that happened to be present. The adaptive cycle and the successional forces would begin anew. The future of this ecosystem would become unpredictable, subject to the likely invasions that occur when successionally advanced ecosystems are disrupted.

Invasive species are considered a detriment to the successional trajectory of an ecosystem because they are thought to spread more aggressively or competitively than native plants. However, a team of researchers looked at the seed dispersal rates of 360 native and 51 invasive plant species. The team discovered that contrary to their expectations, invasive and native plants had similar seed dispersal rates, indicating that invasive species are not necessarily more aggressive. They concluded that "if introduced species do have higher spread rates, it seems likely that these are driven by differences in post-dispersal processes such as germination, seedling survival, and survival to reproduction."[36] In other words, invasive species, like any other species, spread because they encounter ideal conditions in which to thrive. If we want to reduce the spread of invasive species, then trying to eliminate them won't get us far. We need to address the conditions that enable their spread.

The traditional view of succession is that the endpoint or climax community of plants and animals is predictable based on the composition of the former ecosystem. In our example, even if the redwood trees were cut, pioneer, early, and mid-successional species endemic to that region would colonize the forest. However, as we have seen, systems do not necessarily rebound to their former state, especially if processes governing resource exchange have changed. Although the tendency toward a stable climax has largely been discredited within the ecological literature, the idea still persists that invasives alter the "normal" successional trajectory of an ecosystem.[37]

Every species invasion should be evaluated for how it fits into, rather than detracts from, the successional trajectory of an ecosystem. This is especially important when it comes to eradication, because it is impossible to mow, dig, or spray your way out of the adaptive cycle. Like the farmer constantly battling weeds after tilling the soil, these strategies will require consistent use to maintain effectiveness over time. When herbicides are the preferred method of arresting succession, their long-term, consistent, and increasing use to hold an ecosystem at a desired successional state requires close scrutiny. Herbicides are used to manage invasive species as though they will only be needed once, but succession tells a different story – because invaded ecosystems are unlikely to become less prone to invasion by eradicating invasive species. Only through careful guidance of succession through thoughtfully planned disturbance can we effectively manage successional trajectories and the invasive species that affect them.

Stability, Resilience, and Transformation

The idea that an ecosystem's successional trajectory is unpredictable and that invasive species should be understood in terms of nonlinear succession may seem surprising. Although most people accept the idea that ecological succession does not necessarily move toward a definable endpoint and that ecosystems can have multiple stable states, modern species invasions are still widely treated as beyond the scope of what is considered "normal" change within an ecosystem. C. S. Holling, who applied his thinking about nonlinear ecological dynamics to outbreaks of spruce budworms in eastern Canada spruce-fir forests, found that differing views of nature contributed to different interpretations of observed ecological dynamics, which then led to varying management recommendations in response to, in this case, spruce budworm outbreaks.

Holling described three distinct paradigms within ecology, each with its own understanding of "nature."[38] The first view considers nature to be static, changing only in a linear and predictable way. This belief informs the design and implementation of many modern industries, such as conventional agriculture, forestry, and fisheries management, which tend to treat natural systems as limitless supply depots. This approach generally falls short in accounting for the ecological consequences of their designs and practices like nutrient runoff, erosion, and depletion of fish stocks.

The second view is based on the idea that nature is resilient and that natural systems will perpetually regenerate and "bounce back" to what they were prior to a disturbance, independent of the type, scale, or frequency of disturbance. This view encapsulates the premise that invasive species are out of place in so-called "natural" ecosystems, that eradicating them will solve the "problems" they pose, and that ecosystems will return to "normal."

The third view holds that natural systems are evolutionary and always changing. Their resilience is not inherent, but is subject to the scale and nature of the disturbance – and they may change form completely, becoming unrecognizable, but still be stable systems. This third view is inclusive of most modern research into ecological dynamics that accepts nonlinearity and complexity as fundamental to ecological systems.

The first two views are common in invasion ecology and the broader conservation movement. The first, that nature is stable and predictable, permeates the field of invasion ecology, which often seems to react to the presence of invasive species as a surprising and threatening phenomenon. Rather than focusing on predicting the likelihood of invasion based on changes in disturbance or other ecological factors, much of invasion ecology treats each invasion as a separate event unrelated to larger processes occurring in the invaded ecosystem. As Holling predicted, this has to do with ideological beliefs about "nature."

If "nature" is stable and in balance, then an invasion appears to disrupt a finely tuned state of equilibrium. This belief is prevalent: Invasive species are blamed for disrupting the "fragile balance" of nature by upsetting long-standing evolutionary relationships among organisms in invaded ecosystems. Invasive species are "spreading across our lands and through our waterways, and wreaking havoc with . . . fragile native species and ecosystems."[39] The US Forest Service's "Invasive Species Research Model" is based on the idea that

> native systems are thought to persist around a stable equilibrium state. Although disturbances like fire, drought, etc. can shift systems off this equilibrium, successional processes immediately begin to move them back toward the equilibrium. When exotic species invade, they can alter this dynamic. If the invader is very mild mannered, the system may rapidly resume a new equilibrium state so similar to the original that we do not notice the change. However, if the invader is particularly nasty, it can actually launch a new trajectory that may not reach a new equilibrium for hundreds of years.[40]

This research model is troublesome in many ways, first because it subjectively evaluates whether an invading organism is "mild-mannered" or "nasty." It posits that "nasty" invaders can be so disruptive as to launch a new successional trajectory; as we have seen, successional trajectories are not necessarily predictable, especially when the ecological conditions contributing to succession, such as soil conditions, climate, or nutrient availability, have also changed. This definition also relies on a narrow definition of equilibrium, which assumes that historic ecosystems are in equilibrium but invaded ecosystems are not.

A systems-based view acknowledges that ecological systems are constantly adapting to changes and feedback and are in a state of dynamic equilibrium – meaning the "balance" is always in flux – whether we recognize it as such or not. Ecosystems are open systems, and their components and processes are always shifting, from the hatching of insects on a warm day, to changes in breezes and rainfall, to predator/prey population changes, to larger disturbances such as floods and tornadoes – there is constant modification and reorganization in living systems. Despite these changes, ecological systems maintain their system-ness by exhibiting behaviors of adaptability and wholeness over time. As Bill Mollison writes, "Stability in ecosystems . . . is not the stability of a concrete pylon; it is the process of constant feedback that characterizes such endeavors as riding a bike."[41]

Although ecosystems can have multiple stable states, this does not necessarily mean that we should proceed with "business as usual" assuming that "nature" will adjust and adapt accordingly. The type and scale of disturbances acting upon planetary ecosystems today are unprecedented, and it is uncertain whether future ecosystems will support the diversity of life that exists on the planet today, including our own species. We should be actively working to address the forces engendering the massive ecological changes currently underway. If we view invasive species as "canaries in the coal mine" of ecosystems in transition, then we can utilize their characteristics to understand the underlying ecological dynamics enabling their proliferation and work to transform them from the ground up.

Moving Forward

Although salt cedar is considered one of the worst invasive species in the country, an analysis of the highly modified Colorado River watershed could have predicted that a plant with its characteristics would thrive given the existing riparian conditions – and, in fact, it fills a perfect niche. Salt cedar invites the opportunity to understand and engage in whole systems–based assessments of this critical North American watershed and demonstrates that restoration means so much more than eradicating invasive species – if restoration, in the true sense of the word, is in fact the goal.

It makes sense that large dams were constructed to provide electricity and water for centralized agricultural production, that the Bureau of Reclamation was established to reclaim those waters "wasting" to the sea, and that huge cities and agricultural industries would blossom in the desert – because none of these decisions, none of these policies, none of these developments were systems-based. And yet, the facts remain: Sediments will build, dams will fail, salty water will evaporate and leave saltier soil, desert cities will lack water, and all of the life once supported by the "tamed" river will change. Although salt cedar may seem like an inconsequential concern in the face of the Colorado River drying up within the next decade, it's providing an accurate ecological picture of where the system is at this particular point in time. It will continue to change, and it is likely that in fifty or one hundred years, the plant and animal communities will be different as global warming intensifies, glaciers melt in the Rocky Mountains, and rainfall patterns change.

If we desire a Colorado River ecosystem restored to cottonwood and willow dominance, entire industries, legacies, and ways of thinking will have to change. The dams will have to come down – or at least allow for regular pulses of salt-flushing spring runoff and summer monsoons. Many things that our current paradigm holds dear – massive centralized hydroelectric infrastructure to electrify distant metropolises, irrigated agriculture of annual plants to feed animals in confinement and people in faraway ecosystems – will have to be reimagined.

We need restoration, and we need people who care about the ecosystems of which they are part. We need people engaged in redesigning and reimagining our current reality to one that accounts for and enhances the ecological systems we rely on. In the Colorado River basin, we need to turn the focus from eradicating invasive species like salt cedar to working on the much more prescient issues of providing household water via rainwater collection systems for all of the citizens of Phoenix, Las Vegas, and Los Angeles. We need people designing and installing small-scale solar electric and hot water systems to reduce reliance on electricity generated by large dams. We need to build greywater systems, rainwater cisterns, and composting toilets. We need to learn to plant, propagate, and eat perennial desert foods like mesquite, chia, and prickly pear and decrease consumption of foods produced far away with water from even farther away. We need to depave, cut curbs, and slow, spread, and sink every precious drop of water into the land and fit our lifestyles within the ecological potential of the region.

These are systems-based solutions for systems-based problems. If put into practice on a large scale, over time, the salt load in the Colorado River will decrease, and as it does, so too the prolific growth of salt cedar will decline. Cottonwood and willow, the birds that nest in their branches, and the fish that rest in the shade they cast, will return to proliferate in the Colorado River. Eradicating salt cedar by any means necessary will not bring back any of these things.

CHAPTER 4

A Matter of Time

"It is not the strongest of the species that survives, nor the most intelligent that survives. It is the one that is the most adaptable to change."

— *Charles Darwin*

In 1995, a new invasive species was discovered in a lake outside of Butte, Montana. The body of water where it proliferates, known as the Berkeley Pit, is a defunct open-pit copper mine that is slowly filling with groundwater that leaches heavy metals like arsenic, cadmium, zinc, and aluminum from the mine's 2,000-foot-deep surface. The Berkeley Pit is the largest Superfund site in the United States, and as 2.6 million gallons of groundwater seep into the pit every day, the toxic brew edges ever closer to spilling into the Clark Fork River, which makes its way to the Columbia River and out to the ocean, representing a massive-scale ecological threat to the fisheries and communities in these watersheds.

Though the water in the pit has a pH of 2.5, as acidic as lemon juice and considered inhospitable to life, a novel species has emerged and is slowly but surely changing the lake's toxic brew to one more habitable. An analytic chemist studying the lake noticed a clump of green slime floating in the lake and brought the sample to Dr. Grant Mitman of Montana Tech. Mitman brought his colleagues Don and Andrea Stierle into the lab, and they identified the slime as a colony of *Euglena mutabilis,* a single-celled organism that forms algae-like mats. How *Euglena* came to grow in the Berkeley Pit is unknown, although some attribute its spread to a flock of 350 snow geese that landed (and subsequently died) in the lake several months prior to the discovery of the organism. The geese may have carried spores of the *Euglena* in their feces.

Euglena are among the oldest organisms in the world, having survived since the time when conditions on Earth were not too dissimilar to those found in the Berkeley Pit – ancient acidic oceans were full of heavy metals and other free-floating elements. These unique creatures exhibit traits of both plants and animals: They can photosynthesize and produce their own food, but they also move around in search of food. Their ability to photosynthesize produces oxygen, and they also precipitate iron from watery solutions, which creates stable substrates for other organisms to inhabit.[1] In other words, *Euglena* are ecosystem engineers. They work to improve the environment for themselves and, in so doing, make their surroundings more habitable for other organisms. This type of bioengineering fostered the development of all life on Earth, since the production of oxygen led to the proliferation of aerobic organisms, and over the course of billions of years, the stable iron-based substrates eventually contributed to the formation of land masses out of what was once a planet covered by water. Species like *Euglena* manufactured the current configuration of air, water, and land that makes this planet so uniquely hospitable to life today.[2]

In the Berkeley Pit, the *Euglena* are doing similar work. They thrive in the heavy-metal laden waters of the lake and are removing the iron, zinc, and cadmium out of solution, storing it in their bodies, and rendering the metals biologically inert. When they die, their bodies and the metals they contain are deposited in the sediments at the bottom of the lake. The chemistry of the Berkeley Pit is changing slowly but surely, and research is underway to encourage the proliferation of *Euglena* and similar organisms to biologically treat the water before its spills into critical waterways. Since the discovery of *Euglena mutabilis*, more than forty species of similar microorganisms have been discovered in the lake, several of them new to science. Many of these have found suitable habitat because of the pioneering work of the *Euglena*, and they also serve to neutralize and remediate the pit's toxic water. Species like *Euglena mutabilis* are fascinating from an evolutionary perspective because they exhibit traits that not only confer advantages to their own species, but also create conditions that enable other types of life to thrive.

Microbiologist Lynn Margulis observed that the cell organelles such as mitochondria and ribosomes of the closely related *Euglena gracilis* have different DNA than their nucleus. She hypothesized that these highly specialized organelles were once separate organisms that were incorporated into a larger cellular body through symbiogenisis. Margulis believed that the ancestors of these diminutive organisms were once single-celled organisms swimming in the primordial seas of the young planet and that they found mutual survival advantages in joining forces to live together as a single organism. Had the proto-mitochondria and pre-ribosomes continued as separate organisms, plying the oceans in competition for scarce resources, life as we know it on the planet would never have come to pass. Instead, as Margulis and others since have demonstrated, the cells of plants, animals, birds, and insects (the eukaryotes) also contain evidence of this ancient symbiosis. Margulis wrote "Life did not take over the globe by combat, but by networking."[3]

Margulis's theory of symbiogenesis is more widely accepted than when she first introduced it more than forty years ago, but the biological and social sciences are still awash with the idea that competition is the primary driver in evolution. This includes invasion ecology, which still views invasive species only in the light of competition with native species. Researchers spent months rechecking their work when they found that closely related species of algae (which in Darwinian terms should be in heated competition with each other) were in fact sharing resources and preferred to share habitat conditions with multiple similar species.[4]

Invasive, Infective, Collaborative

Invasion and the development of novel symbiotic relationships are the basis of the evolution of life on Earth. When the first plants moved from the ancient ocean onto newly formed landmasses more than two billion years ago, they invaded a completely plant-free environment. Recent research shows that this invasion process was driven by a hormone called auxin, which originated in bacteria and fungi, and which – once part of a plant's metabolism – allowed for the development of roots and vascular tissue.[5] Without this association, sea-dwelling plants would not have survived exposure to the harsh terrestrial conditions of the young planet. This novel symbiotic relationship between plants, bacteria, and fungi drove invasion, which facilitated the chemical weathering of rock substrates, leading to the development of soil, which in turn created conditions for further diversification of plants and animals.

Again and again, invasion is found at the heart of some of the major evolutionary shifts over the course of our planet's long history. The first photosynthetic organisms, which used the young planet's carbon dioxide–rich atmosphere to power their metabolism and growth, invaded the early ocean ecosystem and eventually displaced the anaerobic bacteria that had thrived there for billions of years. Known as the Great Oxygenation Event, this early invasion caused the extinction of most organisms present on the planet at that time – anaerobic bacteria – as the atmosphere slowly changed to favor those that thrived in the presence of oxygen. Sea-dwelling animals moved onto land on their four stumpy legs, irrevocably transforming the formerly animal-free ecosystem forever. Saber-toothed tigers invaded North America, finches invaded the Galapagos Islands, people invaded every continent on the planet, and *Euglena* invaded the Berkeley Pit to clean up our toxic waste.

In each of these cases, the invasions changed ecosystems, but there is little evidence to support the concept that "invasion" is inherently harmful. In fact, invasion led Darwin to formulate his ideas about natural selection and evolution. Darwin observed thirteen different species of finches on the Galapagos Islands (an additional species has since been discovered), and noted that each species featured a distinct beak shape. Some had long and slender beaks, while others were short and blunt. Darwin noted the feeding habits of these finches and found that beak shape correlated with the kind of food the finches ate. Finches with short, strong beaks ate nuts and seeds off the ground, while those with long, slender beaks survived off of cactus fruits or insects.

Darwin posited that the Galapagos finches originated from a single species of finch that had traveled from continental Ecuador to the islands, and the variations between species arose over time as the birds developed different feeding strategies, and he was right. The fourteen known species of finch now present on the Galapagos are most closely related to the grassquit bird *Tiaris obscura*, which hails from the humid forest edges and scrublands of Venezuela, Colombia, western Ecuador, western and southern Peru, western Bolivia, and northwestern Argentina.[6] Darwin's finches invaded the Galapagos Islands approximately 2.3 million years ago, just after the beginning of the most recent Ice Age, known as the Quaternary glaciation, which entailed massive shifts in the world's terrestrial and aquatic ecosystems. The invading finches altered the structure of plant communities on the islands through spreading seeds, favoring certain plants or insects over others, and providing food for predators like the Galapagos hawk and short-eared owl. Invasion is part of ecological change and the diversification of life on the planet.

Checks and Balances

Invasions are happening faster now than at any point in recent history, a fact that leads to a great deal of concern since invasive species appear to disrupt the "fragile balance" of nature. These species invasions can appear to overwhelm ecosystems to which they are introduced because the novel organisms lack the ecological "checks and balances" such as predators and diseases that keep their populations stable in their region of origin. The rapid pace of international trade and travel means that more organisms than ever before are moving around the world and some are thriving in new environments. This leads to concerns that the new invader will displace species in the ecosystems to which it is introduced, leading to calls for herbicide-based eradication or the release of biological control organisms. However, evolution is at work even in these scenarios, as organisms respond to changes, pressures, and stresses in their environment with every shift in the wind or drop of rain on a leaf. Charles Darwin theorized that evolution occurs as an organism responds to these changes through the process of natural selection. He wrote:

> It may be said that natural selection is daily and hourly scrutinising, throughout the world, every variation, even the slightest; rejecting that which is bad, preserving and adding up all that is good; silently and insensibly working, whenever and wherever opportunity offers, at the improvement of each organic being in relation to its organic and inorganic conditions of life. We see nothing of these slow changes in progress, until the hand of time has marked the long lapses of ages.[7]

In ecosystems where invasive species seem to be running rampant, natural selection is also at work, sometimes in unexpected ways. As I noted in chapter 1, cane toads were introduced in the 1930s to Australia to control the native Australian cane beetles, whose voracious appetites were threatening the progress of the continent's sugar cane industry. The cane toads, which are native to the Americas, had been introduced in Puerto Rican cane fields in the 1920s to control white grubs – an effort considered successful. But when they were introduced in Australia, the toads weren't able to eat the Australian cane beetles because the beetles often live on the tops of the cane plants that the toads could not climb. The cane toads proliferated in their new home, largely due to the presence of toxic alkaloids, known as bufadienolides, in their skin that killed potential predators. In Australia, populations of several species of native snakes and lizards declined from attempting to eat the cane toads.

A recent study, however, found that the omnivorous native Australian blue-tongue skinks that live where the invasive plant mother-of-millions grows in eastern Australia were able to prey upon the cane toads when they spread into the region.[8] The foliage of mother-of-millions contains a similar class of toxic alkaloids to those found in cane toads, and blue-tongue skinks that had eaten these invasive bufadienolide-containing plants developed tolerance to the otherwise deadly compounds – and were able to also prey on cane toads without being poisoned. The authors found that even blue-tongue skinks that live in areas invaded by mother-of-millions but uninvaded by cane toads demonstrate tolerance to the alkaloid, pointing to the novel coevolutionary adaptation strategy conferred by the invasive plant to the native skink. They conclude that research on potential effects of invasion should be broadly framed and incorporate the possibility that ecological and evolutionary interactions may arise in unexpected places and between widely varying organisms – including invasive species.

Such novel adaptations occur between plants as well. Recent research shows that invasion can stimulate the expression of new traits in organisms that can enhance their possibility of survival in the changing conditions. One study looking at how species adapt to invasion by cheat grass found that though native plants from invaded areas had fewer leaves and lower root biomass, they and their offspring possessed traits that suppressed the growth of cheat grass.[9] From an evolutionary perspective, it is advantageous for all species to come into a state of dynamic equilibrium with their surrounding environment. If a certain organism invades an area, takes up all available resources, and displaces all other organisms as it does so, over time, it will run out of the resources required for its survival, whether this is water, food, light, soil organic matter, or other necessities. "Fitting" into the ecosystem enhances an invader's long-term survival prospects.

Biodiversity

Although invasive species can actually foster ecological equilibrium over time, they're cited as the second greatest threat to biodiversity on the planet today.[10] Originally published in a 1998 Princeton University study, this "fact" has since been cited in academic literature more than two thousand times. The Princeton study quantified a series of threats to biodiversity, including habitat destruction (which received the distinction of being the first threat, though this information is often eclipsed by references to invasive species), pollution, overexploitation, and disease. Though this study meticulously describes the numerous ways that species of concern are currently threatened, it made the mistake of isolating invasive species as a causal agent rather than a result of the other named threats. Invasive species are not acting alone to threaten biodiversity. They are not, by themselves, to blame.

Though invasive species are considered among the greatest threats to biodiverse ecosystems, there have been only a few attempts to quantify the amount of land covered by invasive species. One of the few took place in the United States and concluded that invasive species cover 18 million acres in twelve southern states (although only 12,188 acres were actually sampled; the rest of the data was extrapolated by the authors).[11] Although the study intended to show that invasive species were threatening ecosystems because they covered such a large area, the study failed to disclose the amount of land covered by agricultural crops, which would have provided important context for assessing whether and how invasive species should bear the burden of blame for threatening regional biodiversity.

Had it provided this larger context, the study could have found that three plants covered nearly 23 million acres across the same region: corn, soy, and cotton.[12] The irony is that the study recorded the presence of thirty-three invasive species – *thirty more* than the three found in the adjacent agricultural ecosystems, making those acres covered by invasive species *more* biologically diverse than the majority of land currently managed for agricultural purposes in those states. This study claimed to be attempting to quantify perceived threats to biodiversity; turning a blind eye to agriculture and other drivers of habitat destruction (the *primary* threat to biodiversity) leaves out a critical bit of information that should be used to determine how to best proceed with the enhancement and conservation of global biodiversity.

Compared to agricultural settings where noncrop organisms like birds and insects tend to be considered pests, the thirty-three invasive species found by the study are known for their beneficial ecological associations. The researchers found silk tree, a favorite of hummingbirds, bumblebees, and other pollinators, as well as roses and Japanese honeysuckle, also favorites of birds and pollinators. All organisms, including invasive species, require the participation of other organisms to ensure their survival – plants depend on pollination and seed dispersal to survive and spread, and animals need adequate food and habitat. If invasive species are spreading and thriving, then they benefit from, and are benefiting, such ecological associations. One study found that "seeds of many of the most notorious plant invaders are dispersed by animals, mainly birds and mammals," demonstrating that even the world's "worst" invasive species are being used by other organisms in their new habitats.[13]

A 2011 University of Pennsylvania study found that areas densely covered by invasive honeysuckle and privet supported greater populations and diversity of native birds like robins, gray catbirds, and other fruit-eating species that rely on honeysuckle for food than test plots with fewer or no honeysuckles.[14] They found that the honeysuckles may serve as a "main axis" of organization of bird communities in the region – meaning that they have very few other food resources available so their populations are dependent upon the ample fruit resources offered by the shrubs. In this case, invasive species are serving as the foundation of maintaining biodiversity. The researchers also planted native black nightshade in experimental plots with varying densities of honeysuckle and found that more fruit was removed from the black nightshade (increasing its likelihood of dispersal) where honeysuckle densities were highest. The authors hypothesized that the birds were attracted to eat the honeysuckle fruits and also ate the black nightshade at higher rates when they were in the area. Compared to areas with less honeysuckle, the highly invaded sites contributed to improved bird and plant biodiversity.

More recent research shows that singling out invasive species as threats to biodiversity not only misses the larger picture of causal factors contributing to biodiversity loss, but also may be true only at a small scale. A 2011 analysis that collated all experimental and observational studies on the effects of species invasions on biodiversity found that scale was an important factor in determining how species invasions affected native organisms.[15] The authors found that in samples taken from plots less than 100 square meters, invasive species had a negative effect on biodiversity, but that on broader spatial scales, the negative influence of invasive species is unclear, even within the same ecosystem.

Another study found that attempts to eradicate the invasive Himalayan blackberry in Sacramento County, California, led to significant declines in population of the nearly endangered native tricolored blackbird, a Californian Species of Concern. Researchers found that the tricolored blackbird had been using the Himalayan blackberry, as well as patches of other invasive species like Canada thistle and bull thistle, as nest sites and that nest sites in these weedy areas had more successful fledglings than those in the native wetland ecosystems.[16] The authors point out that the success of invasive species–based nest sites is directly related to significant declines in the original nesting habitat of the birds.

> Much of the breeding habitat of Tricolored Blackbird today consists of vegetation that differs from that of its original habitats. Of those colonies observed during the 1930s, *c.* 97% of breeding occurred in the vast deep-water emergent marshes of cattail *Typha* spp. and bulrush *Scirpus* spp. throughout California's Central Valley . . . The preponderance of upland nesting that is found today was not reported during this time and the vast majority of upland substrates used now consist of nonnative plant species that would not have been present in the Californian landscape prior to the arrival of Europeans. Although the historic range of Tricolored Blackbirds has changed little since the 1930s, approximately half of all nesting is now in upland habitats. This apparent shift from wetland to upland is surely due to the loss of 96% of California wetlands over the last 150 years from 1,500,000 ha before European settlement.[17]

The "vast deep-water emergent marshes" of cattail and bulrush of central California, which once covered over 3.5 million acres and supported an incredible diversity of waterfowl and other birds, fish, reptiles, amphibians, mammals, and people have been plowed into monoculture fields of oranges, grapes, tomatoes, almonds, carrots, and cotton. Once numerous native species have suffered massive population declines or extinction due to wide-scale adoption of modern agriculture. The tricolored blackbird has less specialized nesting and feeding needs than some of the now-extinct species, allowing them to successfully transition to the use of alternative habitat resources – in this case, the invasive species that proliferate on the field margins.

Eradication strategies must be carefully analyzed in terms of an invasive's ecological relationships to ensure that biodiversity is not decreased further by their removal. If Himalayan blackberries will be removed, for example, managers need to incorporate a plan for a suitable nesting habitat for the tricolored blackbird. A removal project must be phased to ensure that it can maintain populations of invasive species until the time when native species are once again filling these niches. The bottom line, however, is that eradication of invasive species is not the answer to the biodiversity crisis. Indeed, it may well cause more harm than good.

The Princeton study that named invasive species the second greatest threat to biodiversity and found that habitat destruction is the greatest threat didn't correlate the two phenomena (it could have posited that there is a relationship between habitat destruction and the appearance of invasives). Of course, removing invasive species is more straightforward than attempting to redirect the juggernaut of habitat-destroying industrial agriculture or forestry – with all of the subsidies and lobbyists, embedded manufacturing industries, capital intensive assets, fluctuating markets, and entrenched production and distribution models. The trouble is that *these* systems are the major threats to global biodiversity, and the necessity of addressing them is both more urgent and more deserving of our time, energy, research, and experimentation than this militant focus on the eradication of invasive species.

The Relationship between Invasion and Extinction

Extinction is a hard fact in the complex story of life on Earth. Over 98 percent of the species that have ever lived on the planet, including entire classes of organisms, like the dinosaurs, that were diverse, numerous, and successful in their environments have gone extinct. The planet has experienced at least five, and up to twenty, major extinction events throughout the course of Earth's history.[18] In addition to these major events, life on Earth appears to have a "background" extinction rate of one species out of every million per year. If there are ten million species on Earth (the most current estimate, though some believe it is as high as one hundred million), then it's reasonable to expect that ten species would go extinct every year. Though diversity of life on the planet has increased both because of and in spite of extinctions, the recovery of diversity after historic extinction events has taken anywhere from six million to thirty million years. The loss of any species today is not something to be taken lightly, since an organism's unique niche and attributes will not likely be replaced for an almost unimaginably long time.[19]

Today, we are in the midst of what many scientists consider the sixth great extinction, with rates up to ten thousand times higher than "background" rates. Some calculations estimate that up to twelve species go extinct every day.[20] This latest pulse of extinction is known as the Anthropocene because it is largely human driven. Invasive species are blamed as one of the main drivers of the modern extinction event because their rampant growth appears to displace and leave no room for less hardy organisms. But again, the interplay of extinction and invasion is more complex than a simple causal relationship. In fact, it appears from analysis of historic extinctions that invasion processes co-occur or take place after the extinction to take advantage of new niches. In the majority of recent extinctions, competition from introduced species is only one of many causes. Foremost is the destruction or degradation of habitat, which makes it difficult or impossible for imperiled native species to survive.

In an analysis of how modern extinctions are influenced by invasive species, ecologists Jessica Gurevitch and Dianna Padilla of Stony Brook University, New York, found there is little or no conclusive evidence that invasive species represent a significant causal factor in the majority of species extinctions and that of the 762 species pushed to extinction over the past few hundred years, less than 2 percent were caused by the invasion of alien species.[21] According to John Kartesz of the Biota of North America Program, there has never been a plant species in North America, (including Hawaii) that has been driven to extinction by the introduction of another plant species.[22] These findings demonstrate that "most imperiled species face more than one threat, and it is difficult to disentangle proximate and ultimate causes of decline or interactions between different threats and to evaluate their relative importance. Exotic species might be a primary cause for decline, a contributing factor for a species already in serious trouble, the final nail in the coffin, or merely the bouquet at the funeral."[23]

Although organizations like the World Wildlife Federation blame invasive species for "hundreds" of extinctions, some in-depth analyses of recent extinctions reveal that invasive species might coincide with and contribute to extinction processes, but the novel organisms are not necessarily the primary drivers. For example, the Nile perch is often held responsible for the extinction of several species of fish in Lake Victoria, but populations of native fish likely began to decline from overharvesting starting in the early twentieth century, incipient with the development of railroads, overharvesting with increased human populations in the region, the intensification of agriculture, and the resultant erosion and nutrient accumulation in the lake.[24] Excess sediment and nutrients decreased water quality, resulting in eutrophication and fish die-offs. In turn, these conditions favored the growth of an invasive plant species, the water hyacinth, which is also blamed for the extinction of native fish in the lake because of its prolific growth in the shallow water regions preferred by young fish. Although proliferating water hyacinth may present challenges for certain species of fish, and the predatory nature of Nile perch contributed to declining populations and perhaps even finished them off, these organisms didn't singlehandedly cause the extinction of native fish. If these invasive species were removed from the lake, native fish populations would still be in trouble because of increasing deterioration of water quality and continued overharvesting.

True, the introduction of new species has contributed to extinctions in certain isolated cases. One such example is the introduction of the brown tree snake in Guam. Nearly two thousand miles from continental Asia, Guam's isolation offers a unique glimpse into microevolution, invasion, extinction, and speciation. If organisms can travel the distance and efficiently exploit the available resources on islands, chances are, they can invade a new habitat that will likely have more available niches than where they arrived from. In such places, invasion is rare, but when it happens, it tends to have a significant impact.

However, because the complexity of island ecology decreases as distance from continents increases, islands like Guam are particularly susceptible to extinction events. The few cases where it's clear that invasive species have directly caused extinction have taken place on small, isolated islands like

Guam. In fact, 75 percent of modern recorded extinctions have occurred on islands.[25] If a single species goes extinct on an isolated island, the amount of time that the island has been physically separated from a colonizing land mass and the distance from a colonizing land mass figure heavily in whether or how the island will be repopulated. This is because the food web is less complex and influenced only by the type of species that can fly or swim the distance from their continental points of origin. In this sense, not all invasions are created equal. Because of its isolation, Guam had no endemic predators – no predatory animal was able to fly or swim the distance to establish a successful breeding population – representing a huge open niche for any predatory species that found its way there. In short, Guam's ecosystems were (and still are to some extent) simple and therefore prone to significant alteration, especially once people were able to easily travel and bring organisms with them.

The brown tree snake is native to eastern Indonesia, northern Australia, Papua New Guinea, and several Melanesian islands. Though it preys on birds and small mammals in these ecosystems, it has not caused their extinction. Instead, the snake exists in dynamic equilibrium with its preferred prey and is also preyed upon on larger and more biologically diverse islands. Darwin theorized that organisms hailing from more diverse ecosystems would be more likely to survive and become invasive, because they have adapted to numerous stressors that may not exist in less diverse regions. Ecologists Dov Sax and Jason Fridley recently tested this hypothesis and found that like the brown tree snake, most invasive species hail from ecosystems with a wide diversity of selection pressures that promote a higher likelihood of subsequent adaptive radiations.[26]

The brown tree snake was introduced to this isolated and historically snake-free island in the 1940s – researchers hypothesize it came hidden in cargo from nearby islands where it is native. The brown tree snake preyed upon several species of native bird, reptile, and amphibian and is blamed for the extinction of twelve species of birds native to Guam. By the 1960s, the snakes were found throughout the island, and by 1985, the population of brown tree snakes peaked at an estimated one hundred snakes per hectare, though today their populations have decreased substantially.[27]

The birds that survive on Guam are those that have adapted to evade predation – with camouflaged nesting sites, colored plumage, and feeding at different times of day when the snakes are less likely to be around. Guam's birds, like Darwin's finches, were pushed to diversify by environmental pressures. Evolutionary theory would predict that over time, new species will emerge that are well-adapted to coexist with the snakes, similar to the numerous and diverse bird populations of larger islands and continents where snakes and other predators are important parts of terrestrial ecosystems.

And yet, even in this case, it is difficult to place blame solely on the snakes as the primary or sole cause of extinction. As always, there's a larger context at play. During World War II, Guam was the site of fierce fighting between Japanese and American forces. Military bombing campaigns set fire to the island's southern forests, and forests in the north were cleared for the construction of military installations. After the war, the US military introduced tangan-tangan brush to the island from the air to prevent erosion of the denuded landscape. This was helpful to save the soil from washing to the sea, but it didn't replace the forest ecosystem – or the habitat value that it once served. By 1983, only 160 hectares of original forest remained on the island, and much of the island's landscape was transformed into brushlands or savanna-type ecosystems.[28]

The US Army also spread DDT by hand and by plane throughout Guam as often as once per week in the years immediately following the war and continued through the 1970s, though malathion replaced DDT after DDT was banned.[29] DDE, the breakdown product of DDT, was measured in carcasses of the Vanikoro swiftlet on Guam in 1975, demonstrating that its known propensity to bioaccumulate up the food chain may have been acting on insect-eating birds on Guam.[30] Guam also adopted Green Revolution agricultural technology after the war, including the conversion of land into monoculture crops reliant on pesticides and herbicides. Many of the native insectivorous birds are known to have declined because of overuse of pesticides following the adoption of Green Revolution agricultural technologies after World War II.[31] Most of the bird species now extinct on Guam were forest-dwelling and insect-eating birds.[32] Though the brown tree snake certainly did eat birds, Guam's ecosystems were completely transformed by human activities that did not favor forest-dwelling bird life by the time populations of the snakes peaked in the 1980s.

And though the brown tree snake is held responsible for the extinction of some species of birds, the full picture of bird life on Guam and within the greater region tells a slightly more complex story. One hundred and two species of birds have been recorded on Guam, and seventy-two of these migrate through the area. Thirty are resident species, of which twenty-two are native and eight are introduced. Only two of these species are *endemic* to the island – meaning they are found nowhere else in the world.[33] One of these, the flightless Guam rail, is extinct in the wild on Guam, but is being raised in captivity and has been released on the island of Rota forty miles north of Guam. Only one endemic bird unique to Guam, the Guam flycatcher – an insectivore whose populations are thought to have declined because of pesticide overuse[34] – is known to be globally extinct.

Though the brown tree snake contributed to the declining populations of birds on Guam, its role as a direct causal agent of extinction remains unclear. When it comes to comparing extinction pressures, even if the brown tree snake were directly responsible for the loss of nine species, four others are known to have gone extinct because of overharvesting and habitat degradation before the brown tree snake even arrived on the island – pointing to the fact that these types of pressures play an important role in driving extinction processes on the island.[35] The introduction of brown tree snakes certainly caused changes in bird populations and affected the ecology of Guam in important ways. Yet even in this case, which often serves as a compelling example of the ways that species invasions can lead to extinctions, the co-occurring processes contributing to the advent of extinction cannot be ignored. The story of Guam's historic ecological degradation is rarely told when discussing the apparent threats posed by the brown tree snake, but should be openly discussed and acknowledged as an underlying driver of the unfortunate decline in avian diversity on the island.

Although even minimal analysis will reveal that the "worst" invasive species are not the true drivers of extinction as much as co-occurring phenomena, this "blame game" powerfully drives eradication campaigns in less isolated ecosystems, as well. In the Pacific Northwest, for example, the northern spotted owl is critically endangered, and the US Forest Service has named the barred owl as the greatest threat to its survival. Though both owls are native to the Pacific Northwest, the barred owls are considered invasive because they compete with spotted owls for nesting sites and food resources – even though there is still uncertainty regarding the actual effects of the barred owl presence on populations of the spotted owl.[36] However, in 2013, the US Fish and Wildlife Service implemented the Northern Spotted Owl Recovery Plan, which includes a pilot project to use lethal and nonlethal methods of removing barred owls in four locations to determine how populations of spotted owls respond.[37]

But again, resource competition from barred owls is probably a nail in the coffin for the spotted owl more than anything else since the spotted owl's habitat has been decimated in the region. Spotted owls are cavity nesters; they require large trees with bulky branches to build their nests. They feed on tree voles and flying squirrels that inhabit the upper canopies of old growth forests. Barred owls eat a more varied diet and make their nests in younger trees with smaller branches and have thus found abundant habitat in the clear-cut forests of the Pacific Northwest.

Interestingly enough, the two closely related owl species are interbreeding. An article in *Smithsonian* magazine reports that this practice is "diluting" the genes of the spotted owl.[38] But genes are not "diluted" by the addition of other genes like coffee is diluted by the addition of water. Interbreeding selects for the best traits the parents have to offer and creates offspring that are more fit to survive the new conditions. Other species, including polar bears and grizzly bears, are interbreeding, too. Grizzly bears are more widely adapted in their feeding strategies and habitat needs than polar bears, which are dependent on ice cover for hunting seals and walruses. As climate change intensifies, less Arctic ice is a reasonable prediction, and the offspring of the interbred bears may be better suited to survive changing conditions.

Of course, if we lose either the spotted owl or the polar bear (or both), we'll be incalculably poorer for it. We *should* fight to prevent extinctions when we reasonably can, and we *should* mourn the loss of those that have already gone extinct. That said, we should also fight – and mourn – the loss of Arctic sea ice and old growth forests that have been converted into agriculture-style plantations by the modern forest product industry. Shooting barred owls will not touch the underlying processes contributing to the extinction of spotted owls, just as eliminating grizzly bears will not bring back the Arctic sea ice that polar bears are adapted to hunt on. Gurevitch and Padilla write that "if invasives are not a primary cause of extinction or major contributors to declines of species (locally or globally) but are instead merely correlated with other problems, the resources and efforts devoted to removing exotics might be better focused on more effective means to preserve threatened species."[39]

What does that look like? In the case of the spotted owl, we need to create and protect the habitat it needs. This means embracing forestry practices that enhance advanced successional or old growth ecosystems – read: no clear-cuts. To start, we need to favor small, cooperatively owned woodlots that practice conscientious management for fuel, fiber, food, and medicine rather than multinational corporations that treat forest ecosystems as wood and pulp farms. The results of this kind of alternative forest management strategy will be manifold: clear flowing streams, sequestered carbon, strengthened regional economies, and habitat for diverse species – including the spotted owl.

Extinction is a serious matter. Although life has continued and diversified in spite of past extinctions, we can't take the current extinctions lightly – especially since we've coevolved with and rely on Earth's biota. We need a paradigm shift to save species from extinction, and it starts with the creative reimagining and redesigning of the ways we meet our needs.

Climate Change and Carbon Sequestration

Current projections indicate that up to 37 percent of existing species may be "committed to extinction" by 2050 due to the effects of climate change.[40] The past twelve thousand years have been relatively stable climatically, allowing for the development of agriculture and complex civilizations, but it's come at a cost: massive amounts of carbon released from the soil through tillage-based agriculture and forest clearing, as well as the combustion of ancient carbon stores in the form of coal, oil, and natural gas. Though the levels of carbon dioxide and methane in the atmosphere are not unprecedented in Earth's history, they are higher than they have ever been in at least the last 650,000 years and may potentially reach higher levels than any time within the last twenty million years.[41]

The International Panel on Climate Change (IPCC), a consortium of thousands of scientists from around the world, predicts that increased carbon dioxide and other greenhouse gases could increase global temperatures by 5 degrees Celsius (9 degrees Fahrenheit) by the year 2100.[42] The last time global temperatures were this high was during the Paleocene-Eocene Thermal Maximum (PETM) over fifty-five million years ago, when tropical and subtropical forests covered nearly the entire globe. Humans were not yet present on the planet, only our primate ancestors were, so we have no model for how we might adapt to such extremes. The first effects – more intense storms, droughts, floods, decreased snowpack and melting glaciers, higher global surface temperatures – are already upon us. The environmental journalist Dianne Dumanoski refers to these coming changes as "the end of the long summer" in her book of the same name. She writes:

> Of all the hurdles that lie ahead, the most formidable may simply be to recognize that the world has changed fundamentally and that we must prepare to meet a future that may bear little resemblance to what we have come to expect. There is no guarantee that the centuries ahead will unfold as some recognizable version of life we live today. The long view of the human journey testifies that no one living on a dynamic, changeable Earth is entitled to expect the secure, comfortable life the rich of the world have come to take for granted . . . Over the past 11,700 years, humans have enjoyed the climatic blessings of the long summer, an interglacial period that has been extraordinarily lengthy and tranquil. This rare interlude in climate history, this special landscape of possibility, has allowed humans over millennia to construct a global civilization. Now, as we face its end, tomorrow's possibilities may be altogether different. [43]

Changes in "tomorrow's possibilities" are certain for us and have important ramifications for changes in ecosystem structure that we observe today, including the presence of invasive species. Biologist Gerald Rehfeldt of the US Forest Service found that as climate change intensifies, at least in the western United States where his models are based, 47 percent of plant communities will transform from their current composition into unpredictable assemblages of species.[44] Increased carbon dioxide may also lead to slower rates of succession and allow for early successional species (including many invasive species) to persist in communities for longer than they do currently.[45]

Some invasive plant species are more tolerant of carbon dioxide than native species. In one study looking at four different *Rubus* species, two native and two invasive, the invasive species were able to photosynthesize (and thus sequester carbon) for longer periods of the day, using less water and nitrogen.[46] Another study from China showed that three invasive species (*Mikania micrantha*, *Wedelia trilobata*, and *Ipomoea cairica*) demonstrated a 67.1 percent increase in photosynthetic capacity compared to a 28.4 percent increase for co-occurring native species when grown in a high-carbon-dioxide environment.[47] In this context, not only did the invasive species photosynthesize more and grow faster, but they also captured more carbon than native species grown in the same conditions.

The ability for certain species to thrive in a carbon-rich atmosphere may initiate a negative feedback loop that reduces carbon dioxide in the atmosphere. During the PETM, when both atmospheric carbon dioxide levels and global temperatures were very high (similar to what they are predicted to be within the next century), a small water fern called azolla bloomed and spread throughout the Arctic Ocean. Azolla is a unique plant in many ways, one of them being that it is part of an ancient symbiotic relationship with a nitrogen-fixing bacteria of the genus *Anabaena*. The plant can fix up to 1,000 pounds of nitrogen per acre every year and is used today in organic rice cultivation for this purpose. During the PETM, this fixed nitrogen provided necessary nutrients for the azolla, which in turn used the ample atmospheric carbon dioxide as fuel for growth of its extensive populations. Azolla can fix 6,000 kg of carbon per acre per year floating on water because of its nitrogen-fixing abilities. The unique geologic character and warm temperatures of the era meant that the Arctic Ocean had a lot of freshwater floating on its surface, to the point where its salinity matched the salinity tolerance of azolla. Over the course of approximately eight hundred thousand years, atmospheric carbon levels dwindled from 3,500 ppm to 650 ppm, due in part to the rampant growth and carbon sequestering abilities of this aquatic plant.[48] As the azolla grew and died, the dead plants floated to the seafloor, full of atmospheric carbon. Today these layers of dead azolla are known as a "petroleum source rock" that may extend beneath the entire Arctic Ocean.[49] Many researchers speculate that the growth of this plant was directly responsible for creating the more moderate modern climate.

Considering the role of a diminutive plant in shaping planetary climate history has important implications for understanding invasive species, many of which are more tolerant to increased levels of carbon dioxide. Is the climate of today, with atmospheric carbon dioxide levels increasing toward levels unprecedented in human history, acting as a selective force for plants that can withstand and thrive in these conditions?

Ecosystem Conversion

Hailing from eastern Asia, the emerald ash borer is considered one of the most destructive pests in North American forests. These diminutive beetles are thriving in ash trees throughout Michigan, Colorado, Connecticut, Georgia, Illinois, Indiana, Iowa, Kansas, Kentucky, Maryland, Massachusetts, Minnesota, Missouri, New Hampshire, New York, North Carolina, Ohio, Pennsylvania, Tennessee, Virginia, West Virginia, Wisconsin, Ontario, and Quebec. Emerald ash borer larvae tunnel extensively below the bark surface, disrupting a tree's ability to transport nutrients and water. This disruption inhibits its growth and can lead to death within two to four years after infestation. To date, an estimated two hundred million trees have been killed as a result of the beetle infestation. In attempts to curb the spread of the beetle, nursery stock is subject to quarantines, and all firewood must be kiln dried or fumigated with methyl bromide before interstate sales.

Because ash trees line sidewalks in residential neighborhoods and comprise large swaths of forests, pesticide application policies are often designed to save particularly valuable and visible ash trees in public places. Common preemptive treatment includes the injection of systemic insecticides into ash trees in areas where the borers are known to live. In order to be effective, these injections must continue every year for at least twenty years to ensure the borers are unable to establish populations. Bayer, which manufacturers the systemic insecticide imidacloprid under the trade name TreeAge, markets the product to homeowners in regions afflicted by emerald ash borer, and some municipalities also use the product to a limited degree. Imidacloprid is a neonicotinoid insecticide – the class of insecticides implicated in honeybee colony collapse disorder. Though municipalities have yet to adopt formal policies of insecticide-based eradication of the borer, imidacloprid is still used in residences, cities, and counties throughout the region affected by the insect, and several agencies are weighing the risks and benefits of its widespread application to combat increasing ash borer populations.

The trouble is that ash borers take up residence in trees stressed by drought or pollution. That pests tend to attack plants experiencing stress, whether from drought, deprivation of essential nutrients, or physical damage, is well-researched in the field of organic agriculture.[50] For the last decade, the northeastern United States has had below average rainfall, which leads to long-term drought stress for plants adapted to wetter conditions, including the ash trees. A healthy ash tree should be able to produce enough sap to push the ash borer larvae out of the cambium, but unhealthy trees use their sap resources to maintain basic circulation, allowing the beetles to colonize. The emerald ash borer is not so much an invasive species as a species that is taking advantage of changing ecosystems.

Instead of wasting valuable resources and adding further toxic chemicals to the environment in futile attempts to eliminate the beetle, land managers in the affected regions should start experimenting with alternative species that are likely to thrive in the warmer, drier conditions that are likely to emerge as the effects of climate change intensify. Look to those ecosystems three climate zones in all directions, propagate trees (and other plants) from there, and plant them. Discover species from analogous climates from around the world, and do the same. We can mitigate the effects of climate change by acting with full acknowledgment that it is happening and by taking meaningful action to enhance diversity and resilience of terrestrial ecosystems by using all available options, including introducing organisms from other places to discover what survives.

Habitat Connectivity and Assisted Migration

Most climate scientists predict that the climate is changing and will continue to change, rapidly, and many economically and ecologically important species won't be able to keep pace either by adapting to their new conditions or by "moving" to a new, more suitable habitat. A recent study of historic migration patterns of 540 species of terrestrial vertebrates, including amphibians, birds, reptiles, and mammals, found that on average these organisms have taken one million years to adapt to a temperature change of 1 degree Celsius. If global temperatures rise an average of 4 degrees Celsius over the next one hundred years as predicted by the IPCC, these organisms will need to adapt to changing conditions ten thousand times faster than they have done in response to historic changes in climate.[51]

Organisms are already moving to new habitats in response to climate change even though this is neither rapid nor easy. Plants mostly rely on wind, water, and animals to facilitate their spread; pollen records indicate that trees moved an average of 50–2,000 m/y^{-1} as the glaciers retreated at the end of the last ice age.[52] The California yellow lupine has recently been found growing on sand dunes two hundred miles north of its "native" range – perhaps it is following its preferred temperature and moisture regimes as it shifts due to climate change – and although this perennial nitrogen-fixing shrub is stabilizing the dunes, attracting pollinators, and improving the fertility of the sandy soil, it is considered an invasive species and is subject to eradication measures. Animals, of course, are more mobile than plants, but realistically they only travel as far as they need to obtain resources they need for their survival – the type of food they eat, their nesting habitat, and so on. Attempts to keep plants and animals where they are "supposed" to be fail to take climate change into consideration. This rapid shift in temperature gradients is forcing species throughout the world to either perish or migrate in elevation or latitude. Plant and animal species that are biologically capable of moving to new climate zones are doing so in order to survive.

A 2014 Cornell University study found that four species of birds inhabiting the slopes of remote mountains in Papua New Guinea have moved their range an average of 400 feet up slope over the past fifty years.[53] Temperatures in the region have increased an average of 0.7 degrees Celsius during this time, leading to changes in food availability. If warming continues as predicted, the habitat required by four species that have moved from the slopes to the summits of mountains will literally disappear. Comparable resources may be found further north, perhaps in Taiwan or China; if the birds can make it there, and there are adequate niches to foster new ecological relationships, then there is a chance that these birds will survive. But it's also possible that their preferred habitats and food resources will disappear completely, and the birds will have nowhere to go.

In order to mitigate the possibility of mass extinction, we need to ensure that these slow-moving or geographically isolated organisms can get to the habitats that suit them best. This means designing and implementing habitat connectivity for ease of migration. Most development, whether planned or unplanned, ignores the importance of habitat for any organism other than people. Suburbs encroach into the small bits of remaining habitat, and diverse ecosystems are converted into commodity-production zones. Imagine you are a lizard endemic to the coast of Baja California, Mexico.

As the climate warms, your ideal habitat shifts northward by two hundred miles to Santa Barbara. Over the course of several generations, you and your progeny may be able to make the long journey to a more suitable locale, but in the course of doing so, you have to cross the freeways and urban landscapes of Tijuana, San Diego, Oceanside, San Clemente, Irvine, Long Beach, Los Angeles, Santa Monica, and Oxnard. Chances are, you and your family are not going to make it.

On a regional scale, prioritizing habitat connectivity means envisioning a continuous corridor of "wild" spaces that butterflies, birds, lizards, frogs, skunks, bobcats, bears, and more can easily move through. These spaces must be large enough to provide food and water resources, nesting habitat, and places to hide. These corridors must also be substantial enough for seeds of trees, shrubs, grasses, and forbs to take root and propagate as they make their slow way to cooler climates over the course of multiple generations. Implementing this is no easy task. It requires creativity and a comprehensive redesign of our means of procuring food, water, and material resources into more integrated and intensive models of production. Instead of relying on far-flung farms and stores, we must begin to see our homes and communities as integrated parts of the natural world and design them with the health and survival of all other organisms in mind.

As we grapple with the probability of losing 37 percent of the species on the planet in the coming years, and the likelihood that at least half of global ecosystems are going to transform in their structure, function, and species assemblages in unknown ways as climate change intensifies, the question of what species we want to save, and how, becomes poignant. We have an opportunity to save some of the species that may otherwise be lost, and one of the important contributions we could make is to plan assisted migrations. This process will require thoughtful and protracted observation, since it's necessary that the move be beneficial to both the moving organisms and the ecosystems that receive them.

Assisted migration is a controversial subject, largely due to concerns about introducing species that will become invasive. In an assessment of invasion risk posed to ecosystems by assisted migration, ecologists Jillian Mueller and Jessica Hellman found that the risk that introduced organisms would displace other species is low, especially when species are moved around within continents.[54] As we have seen, the dynamics of invasion are complicated, especially over time, and invasions have rarely directly caused extinctions. Subsequent increases in speciation and adaptive responses to competition have increased biological diversity.[55] In New Zealand, for example, there are currently 4,000 species of plant flora, up from 2,000 species before European contact. Hawaii is now home to 2,300 plant species, up from 1,300. Even Easter Island doubled the number of plant species it supports from 50 to 111 since the time of European colonization.[56] Although biodiversity relates not only to the overall numbers of species but also to their functional relationships, given the uncertain climate future, a greater number of organisms living in diverse regions of the world imparts a statistical likelihood of survival for a particular species.

Historic species introductions have broadly increased populations of some organisms that could otherwise be vulnerable to extinction. The Monterey pine, for example, is endemic to coastal California, occurring only in three small groves located within 125 miles of each other. The limited range and relatively small population make this species vulnerable to extinction from the effects of a changing climate, including increased storm frequency, sea-level rise, and intensifying pest and disease pressure. However, Monterey pine was introduced to New Zealand in the early 1900s, where it thrived. Today, the tree is one of the most important timber trees in the country, and is widely planted.[57] If populations of Monterey pine in California were to decline, those growing in New Zealand could provide important genetic material for breeding and reintroduction.

In northern California, completion of the Shasta Dam in 1945 decimated populations of endemic salmon that once inhabited the McCloud River. The depletion of fish stocks also heightened tensions with Winnemum Wintu tribal members, who had stewarded salmon stocks in the region for generations and relied upon them for food and cultural values. Prior to the construction of the dam, a salmon hatchery was also established, and hatchery workers sent eggs of the endemic strain of Chinook salmon to stock rivers all over the world. In New Zealand's Rakaia River, populations of the salmon expanded and thrived. In 2010, a delegation of Winnemum Wintu tribal members traveled to New Zealand to take part in a salmon ceremony, and they also returned with salmon fry to rear in their own hatchery in northeastern California. Today, they are working with dam operators to establish water flows compatible with preserving the salmon, which have not been present in the river for over sixty years. Although the introduction of Californian salmon to New Zealand seems unorthodox today, this action served to ensure the continuation of a unique species, and it is a good example of the types of actions that may be necessary to save species from the multiple pressures imposed by climate change and habitat destruction.

If climate change is as abrupt and severe as predicted, assisted migration should be taken seriously. Plants and animals can be moved from their current locales into areas where the future climate is predicted to be most similar to the conditions in which they thrive. Some forward-thinking activists are already doing this. The group Torreya Guardians transplanted seedlings of the endangered conifer *Torreya taxifolia* from its limited native range in Florida to the mountains of North Carolina, where climate change models predict the existence of suitable future habitat for the species. Oliver Kellhammer, a British Columbian ecologist, planted an experimental forest of dawn redwoods, gingko, and windmill palms – trees that once proliferated in western Canada when temperatures on the planet were 5 degrees Celsius warmer than they are today. Perhaps his experimental plot will serve as a valuable repository of genetic information for post–climate change emergent ecosystems.

Though these and other introductions are changing up the ecological profiles of what people today consider native, acknowledging the real and advancing threats of climate change should stimulate unconventional approaches to conserving and enhancing biodiversity. Although assisted migration is still debated in academic circles, and therefore not widely practiced as a biodiversity conservation strategy, the practice is gaining traction as more people begin to piece together the evidence and come to terms with the scale of systems change of which we are part. Professor Chris Thomas of the University of York writes that

> the argument that translocations will create "unnatural" communities is not particularly relevant in the world today. A philosophy of conserving the composition of biological communities as they are, or restoring them to some specified (or imagined) historical state, sits uneasily with the reality of environmental and biological change . . . Rapid climate change sets up disequilibria between distributions and climate that may take centuries or even millennia to stabilize;

hence, dynamic changes in the distribution of species is inevitable for a substantial period, even if there were to be no further warming (unrealistic as that is).[58]

Invasion as *Unassisted* Migration

Some invasive species have benefited from assisted migration, a move that may be saving the species. Four species of Asian carp, including silver, big-head, black, and grass carp, are critically endangered in their native habitat of the Yangtze River basin in China. Between 1997 and 2005, populations of larvae and fry of these fish experienced a thirty-seven-thousand-fold decline in their abundance.[59] These fish are important parts of Chinese cuisine, literature, and folklore – one of the primary fish food sources, both from wild populations and from fish produced in aquaculture systems.

Asian carp aquaculture is described as early as the fifth century BCE in Fan Lee's *Chinese Fish Culture Classic*. In this text, the author describes how, through the cultivation of carp, he generated a vast income, worthy of envy by a king, who questioned him directly about the foundations of his financial success.

> "You live in a very expensive house, and you have accumulated millions. What is the secret?' Whereupon Fan Lee responded: "Here are five ways of making a living, the foremost of which is in aquatic husbandry, by which I mean fish culture. You construct a pond. . . . In the pond you build nine islands. Place into the pond plenty of aquatic plants. . . . Then collect twenty gravid carp. . . . By the second moon of the next year . . . the total harvest can render a cash value of 1,250,000 coins. The following year, . . . the take will amount to 5,150,000 coins. In one more year, the increase in income is countless." [60]

Asian carp were introduced to the rivers in the midwestern United States in the early 1800s as a game fish and later used for algae and aquatic vegetation management in the Mississippi River, where fertilizer and sewage runoff led to vegetation blooms, on which Asian carp feed voraciously. In the Mississippi River watershed today, these four species of Asian carp are considered noxious invasive species because their feeding habits intersect with those of native and sport fish including walleyes and perch and, interestingly enough, introduced salmon. These carp also startle easily, and when anglers motor through water where they swim, they can leap 8–10 feet into the air. The Federal Register on Injurious Wildlife Species warns that Asian carp represent a public health hazard because boaters have been knocked unconscious and been bruised and bloodied by these giant fish.[61]

Fisheries biologists and land managers fear the Asian carp may soon cross into the Great Lakes via the Chicago Sanitary and Ship Canal, which artificially connects the Mississippi River and the lakes that lie to the north. Of course, the ecology of the lakes has been transformed by species introductions and artificial changes in water flow since the beginning of the industrial revolution. Brian Icke, a fisheries biologist who studies the carp in both the Mississippi and Yangtze Rivers, writes:

> We are trying to keep invasive Chinese carps out of the Great Lakes, to protect an invasive (yet purposefully stocked) Pacific salmon fishery, which was stocked as a management tool to control hyper-abundant alewifes, another invasive fish species, because the native piscivore, the Lake Trout, was nearly wiped out by another invasive species, the sea lamprey, because people built the Welland Canal around Niagara Falls to promote intercontinental shipping deep into the Great Lakes basin.[62]

The invasion of Asian carp is only the latest in a long line of ecological upheavals resulting from industrial and agricultural development on the shores of the Great Lakes and within the Mississippi River watershed. The region's waterways are terribly polluted by agricultural runoff, as well as from steel mills and petroleum coke refineries, fertilizer and pesticide manufacturers, and plastic and battery factories that line the shores of Lake Michigan. Despite the ecological degradation and wave after wave of invasions in the region, only one species of fish has gone extinct – the black cisco, a salmonid whose populations were overharvested by the 1940s, resulting in its extinction by 1960. Seven other morphologically distinct ciscoes are still present in the lakes despite the pressures presented by overfishing and industrial pollution.

Recent genetic analysis of the taxa also demonstrates that rather than being separate species, these eight fish are genetically identical, and their differing physical characteristics are not the result of evolutionary divergence.[63] What were formerly considered eight distinct species are now known to be one species with eight different physical appearances. Advances in DNA analysis can complicate our understanding of biodiversity loss in cases like this – because in a case like this, biodiversity is not truly lost; it's a revision of taxonomic classification. This technology has also fueled the discovery of "cryptic species" – species that are physically similar to each other, but not genetically related. This advance has the potential to significantly *increase* indexes of planetary biological diversity.[64]

Although Asian carp have spread throughout the Mississippi River, and although they abound in many of its tributaries, they have not driven any other species of fish to extinction. Still, there is widespread concern. Federal, state, and local agencies have spent hundreds of millions of dollars on efforts to ensure Asian carp don't make their way into the Great Lakes, but so far the money has been used to deploy largely ineffective techniques. The US Army Corps of Engineers constructed submerged electric fences designed to shock and kill the fish as they approach the lake entrance, which has been plagued with maintenance issues since its completion. In 2009, when the fence was under repair, the Illinois Department of Natural Resources applied 2,200 gallons of rotenone (at a cost of $3 million), a naturally derived poison that affects all gilled creatures, in an attempt to halt the carp's northward spread. Among the tens of thousands of dead fish, researchers found one Asian carp.[65] Other natural resource managers suggest the use of strobe lights or bubble and sound barriers to deter the northward expansion of the carp populations. Researchers at Auburn University are working on developing genetically engineered carp that will produce only male offspring when they mate with their "wild" invasive counterparts.

And yet, these four species of concern are also dying out in their native range. Is it possible to conserve rather than demonize them in an ecosystem where they are clearly thriving? And if their populations are truly threatening other species, can we devise more ecologically appropriate management strategies? Given the realities of climate change and the fragility of our food production and distribution systems, it would be wiser to think of Asian carp in terms of their multiple potential yields. These fish, which can grow up to 100 pounds, have been a dietary mainstay in their native China for thousands of years and literally leap out of the water into boats as they pass by. These attributes make for easy and exciting fishing adventures – at least with the proper safety gear.

Harvesting the Asian carp for food is slowly becoming more popular in the upper midwestern United States, with chefs and anglers experimenting with processing, recipes, and marketing. Louisiana chef Phillippe Parola is working to develop Asian carp as the basis of a viable commercial fishery that will be sold in restaurants throughout the country under the name "Silverfin." He believes that "fish translates to one thing: food. . . . It's one of the greatest natural resources we have."[66] Asian carp could also be harvested for use as natural fertilizer to help transition the millions of acres of conventionally farmed corn and soy – whose fertilizer runoff contributes to algae blooms that feed the hungry fish in the first place – to organic methods of soil fertility building.

Over time, the riparian ecosystems of Indiana, Illinois, and wherever else the carp are present would transition to lower nutrient, higher oxygen conditions. Installing nutrient-harvesting hedgerows, shelterbelts, and wildlife corridors could provide the shade and cooler water temperatures preferred by native fish. Converting land from monoculture crops into riparian floodplain habitat would improve the health of native streamside vegetation by allowing the seasonal pulses of floodwaters to strengthen and over time enhance plant and animal diversity in stream channels. Achieving these multifaceted and long-term restoration outcomes would ensure the return of viable populations of native fish in these rivers. But poisoning, shocking, and shining strobe lights on Asian carp will not.

In the meantime, Fan Lee's 2,500-year-old advice to create aquaculture systems is as relevant today as it was in his time and could be as profitable. Carp farming is still widely practiced in China because it is an efficient use of resources for the production of protein. In *Permaculture: A Designer's Manual*, author Bill Mollison claims that aquaculture systems can expect four to twenty times the yield as an adjoining piece of land.[67] This is due to the ease of nutrient transfer and dispersal in an aquatic environment, as well as the relative consistency of climatic conditions. Water holds a more constant temperature and liquid fertilizers (especially in the form of fish and other manures) can disperse easily throughout it, compared to land-based agriculture where fertilizer requires careful and timely application to be effectively taken up by plants. Aquatic polycultures producing Asian carp as a protein crop could replace some of the meat produced by cows, pigs, and chickens raised in confinement and make available the millions of acres of corn and soy used to feed them for other uses. Asian carp–based aquatic polycultures could also support numerous associated yields, such as duck meat and eggs, mulberries, wapato, bamboo, water lilies, and even rice and/or wild rice.

It is time to start thinking about the species that are thriving in new environments as allies in a quest to more thoughtfully steward our local ecosystems. The $80 million allocated to "control" Asian carp with electric shock fences, bubble strobe barriers, and rotenone could be used instead to support the development of land-based aquaculture systems, small cooperative fish canneries, and fish fertilizer manufacturing facilities. These businesses would also manage the populations of carp by making use of them, turning their profligate growth into food and long-term, local employment opportunities in the regions where the fish proliferate.

CHAPTER 5

Problems into Solutions

"Nature abhors a vacuum."

—*PARMENIDES, 485 BC*

Ecosystems are always transforming, although climate change and other human activities have led to unprecedented transformations. Despite this, life continues to proliferate, thanks in no small way to Gaia – the vast collection of processes that transform matter and energy and without fail continue to produce the conditions suitable for life on the planet. Consider how the air we breathe is the perfect mixture of oxygen and nitrogen, finely calibrated for multiple billions of breaths. The water we drink is the same water that has always been on the planet, the only water that has ever existed, constantly recycled and purified through living systems over the course of billions of years. Most of the planet's seven billion people, and countless other species of animals, are able to eat every day from the bounty of the world's soils. We walk, run, and jump along Earth's surface, free from the mountains of waste we create, thanks to the work of trillions of microorganisms that live on our skin, in our gut, and in the soil. For the most part, these ecological services are so embedded in our understanding of reality that we barely notice them. It's so easy to take them for granted. But at this point in human history, it's necessary to take a hard look at how our actions are affecting these valuable ecosystem services. The results appear rather grim.

In 2005, a consortium of more than one thousand natural and social scientists compiled the wide array of functions and services provided to people by living systems into a document called the Millennium Ecosystem Assessment (MA). This assessment accounted for all major types of terrestrial and aquatic ecosystems, from the alpine tundra to tropical coral reefs, and found that 60 percent of the twenty-four identified ecosystem services are being actively degraded by development, agriculture, forestry, resource extraction, as well as increasing both poverty and affluence. Not surprisingly, the MA found that though "the changes that have been made to ecosystems have contributed to substantial net gains in human well-being and economic development . . . these gains have been achieved at growing costs in the form of the degradation of many ecosystem services . . . and . . . these problems, unless addressed, will substantially diminish the benefits that future generations obtain from ecosystems."[1]

This sobering assessment calls for a thorough reanalysis and redesign of how we make a living from the land. We can no longer design systems that dissociate us from our ecological home. We must become accountable for our needs and our yields, the ecological inputs and outputs that support human life and endeavors. Electricity does not come from the flip of a switch, food from grocery stores, nor gas from gas stations. The ecological reality behind all of these parts of our daily lives needs reckoning, as historic and ongoing severing of interrelationships are causing ecosystems to change in the structure, function, and life-supporting services they offer – and not in ways that will help our species, or countless others, thrive, or even survive.

And yet we are also letting so many resources go to waste. Wetlands are drained and filled, and their valuable services of filtration and nutrient cycling are largely unutilized, degraded, and ignored. Creeks are channeled into culverts and paved over, perpetually hidden from the sunlight, with wastewater and runoff flowing directly into them. Homes are built with materials that have no connection to the ecosystems that surround them. Food is grown with synthetic fertilizers and pesticides, as though plants need ingredients conceived in a laboratory more than the complexity of life in the soil.

As the frequency and scale of ecological damage has increased, so too have ecosystems shifted to mirror these changes. In the early twentieth century, a new form of ecosystem was defined – known as *ruderal*, from Latin, meaning "of the rubble." Ruderal ecosystems are heavily disturbed, marginal lands in which a limited number or type of species can thrive. Railroad right of ways, roadsides, field edges, and factory effluent pools are ruderal ecosystems that are impacted heavily, "waste" spaces that are overlooked as valuable ecosystems because they do not conform to common conceptions of what defines "natural" environments.

Invasive species are often components of such ruderal ecosystems and are performing valuable services in these contexts. Like the proverbial dandelion in the sidewalk, they are reminders that life persists. The living world pulses and grows beneath our feet, despite – and perhaps because of – the concrete of the human-built environment. The dandelion feeds bees, who feed birds, who feed coyotes, who feed hawks. The food web, truncated as it may be, still exists. Invasive species are in many cases compensatory elements of ecosystems in crisis; ecological damage control whose presence is serving to keep life going on some base level.

Though many studies have attempted to quantify the losses of ecosystem services due to the presence of invasive species, there is little definitive evidence pointing to invasive species as casual factors of such ecological degradation.[2] Even declining populations of native species and increasing numbers of invasive species don't prove that novel organisms lack ecological functionality, especially when considered from the perspective of the ecosystem as a whole.

In a permaculture design, every element serves multiple functions, an approach informed by observation of natural systems. Invasive species, like it or not, are part of nature. Observation will demonstrate that they are serving some ecological purpose – whether it aligns with our cultural sensibilities or not. Close observation of invasive species will demonstrate that their presence is always supporting at least one, and often several important ecosystem services, especially in neglected or poorly designed and stewarded lands. The services offered by invasive species are different from those offered by native species, but their different ecological characteristics are representative of real-time ecological dynamics.

If we are to live within highly functional and thriving ecosystems, we have to design systems that account for, enhance, and contribute to these ecosystem services. It's important to understand what ecosystem services invasive species offer in order to develop a holistic approach to their management. They are often representative of how the system has been compromised, and they provide insight into how to best move forward with restoration strategies that enhance overall ecological function.

Nitrogen

Nitrogen is essential to the vegetative growth of all plants. Some require more than others, especially the annual plants we're accustomed to eating like broccoli and corn, but all plants use nitrogen to fuel their basic growth processes. Seventy percent of the air that we breathe is nitrogen, but in a form (N_2) that is not biologically available to plants, which mostly use nitrogen in the form of nitrate (NO_3) and ammonium (NH_4). Considering how important nitrogen is to plant growth, it is surprising that it is naturally made available through relatively few ways. Natural nitrogen fixation occurs through lightning, lichen, and the action of certain bacteria and fungi. The early-twentieth-century discovery of the Haber-Bosch process, which burns air in the presence of a catalyst – forcing atmospheric nitrogen to precipitate into a powder – revolutionized agriculture because it made synthesized nitrogen available in large quantities to fuel the growth of crop plants.

In organic nitrogen fixation, bacteria in the genus *Rhizobium* and actinomycetes (in the genus *Frankia*) form nodules on the roots of certain plants. In exchange for plant sugars, these bacteria convert atmospheric nitrogen into nitrate and pump it into the roots, where it is used for growth. Most of the plants that benefit from the relationship with the *Rhizobium* are leguminous, belonging to the family Fabaceae, including peas and beans, as well as clover, black locust, mesquite, Siberian pea shrub, Scotch broom, vetch, astragalus, acacia, and many more. Plants that partner with the actinomycete *Frankia* are from diverse botanical families and include alders, ceanothus, myrtles, silverberries, Australian pine, and cascara sagrada.

Although these plants use the nitrogen for their own growth, they also fix more than they can use. The excess is deposited in the soil and circulates throughout the ecosystem. Many organic farmers exploit nitrogen fixation for healthy soils by using cover crops and green manures. They incorporate nitrogen-fixing leguminous plants into the crop rotation, allow them to grow until the plants have fixed the greatest amount of nitrogen (at the peak of flowering), and then mow and till them into the soil. Timing is important because after peak flowering, the leguminous plants start using up some of the nitrogen from the nodules to form their nutrient-dense seeds – what we call beans.

Permaculturists frequently design systems to include perennial nitrogen fixers, which they grow for soil fertility much as organic farmers do, but manage the plants by cutting back the leaves and branches to make mulch or compost rather than tilling it into the soil. Pruning stimulates the plants to shed nitrogen nodules as some of the roots die back to compensate for the loss of photosynthetic material (leaves) above. Nearby non-nitrogen-fixing plants benefit from the newly available nitrogen while the pruned leaves and branches make excellent mulch – adding more nitrogen and organic matter to the soil surface as the biomass decomposes. No plowing is required, and the trees and shrubs cut in this way generally resprout vigorously as their roots regrow.

Many invasive species are nitrogen fixers, such as Scotch broom, Russian olive, kudzu, black wattle, sea buckthorn, mimosa, and black locust. Nitrogen fixers tend to thrive in disturbed sites, which is no surprise given their ability to provide for their own primary growth nutrients. Nitrogen fixers change nutrient availability in ecosystems, which in turn further alters the successional trajectory by providing an enriched soil substrate favored by different plants than would have colonized the area had no nitrogen fixation taken place.

In one Hawaiian study, the invasive nitrogen-fixing candleberry contributed 18 kilograms per hectare (kg/ha) of nitrogen per year in an area where nitrogen fixation by native plants amounted to 0.2 kg/ha per year.[3] The native Hawaiian myrtle lacks the nitrogen-fixing actinorhizal symbiont, but the presence of invasive myrtle – with which the native otherwise shares much in common – in these ecosystems adds significantly more nitrogen to the soil. Enriched soils have been shown to attract large populations of invasive earthworms (yes, even worms are considered invasive in certain places). Worm biomass was shown to be up to eight times higher under the invasive candleberry compared to the native myrtle.[4]

There are a few native Hawaiian nitrogen fixers, including the majestic koa and the wiliwili. The koa is prized for its extremely hard, dense, fine-grained wood and is now threatened with extinction. Wiliwili is a light-grained wood, traditionally used for making surfboards and fishnet floats. Both of these species have been overharvested since the rise of modern land and resource use took hold on the islands. Perhaps the loss of these endemic species has led to the proliferation of other nitrogen fixers more suited to the land use and resource management strategies currently practiced in Hawaii.

Nitrogen fixation is a critical ecosystem service that enables the survival of all life on Earth, at least the life that depends on plants, which require the nutrient for their growth. This cycling of nitrogen from atmosphere to soil is accomplished only through the action of microorganisms – either alone or in partnership with plants, lightning, or the energy-intensive Haber-Bosch process. Given the importance of nitrogen to plant growth and development, the demonization of invasive nitrogen-fixing plants is difficult to understand, especially given the alternative of synthetic nitrogen fertilizer production. If invasive nitrogen fixers are proliferating, instead of spraying to eradicate them, we should figure out ways to harness the incredible potential of this critical and freely available ecosystem service.

What would harnessing the potential look like? In tropical areas like Hawaii, native koa and wiliwili, as well as slow-growing but ecologically and economically valuable hardwoods like teak, mahogany, and lama should be planted among invasive nitrogen fixers like *Myrica faya* and *Albizia* – which should be carefully managed and harvested. Permaculture practitioners use the "chop and drop technique" in which vegetative growth is chopped with a machete or billhook and dropped around the base of desired plants as mulch. Managing hardy perennial nitrogen fixers in this way builds soil from two directions. The leafy biomass is rich in nitrogen and builds soil organic matter as it decomposes around the base of the desired plants and the nitrogen-bearing nodules on the roots slough off into the soil as root dieback occurs. Since these invasive nitrogen fixers are perennial, they will resprout and can be managed in this manner for years, serving as natural soil fertility-building nitrogen pumps. Eventually, the valuable native and nonnative hardwood species will shade them out, once they are well established and growing prolifically because of their early boost in soil fertility.

Phosphorous

Like nitrogen, phosphorous is a critical nutrient – plants require it for photosynthesis. Unlike nitrogen, phosphorous is not widely available in the atmosphere or in most soils. Synthetic phosphate fertilizers are manufactured in much the same way as nitrogen-based fertilizers. The rock-based mineral phosphorous is heated in large electric furnaces in the presence of an acid that changes its molecular structure. Rock phosphorous is mined from nonrenewable surface deposits, leading to the inevitable decline of these finite reserves – the estimated "global peak phosphorous" is within the next fifty to one hundred years.[5] The importance of phosphorous to agriculture as we know it cannot be understated, but the raw materials have been grossly mismanaged since the beginning of the Green Revolution.

Phosphorous is a largely immobile element; a plant's roots must be within one-tenth of an inch of a phosphorous molecule in order to use it. Because of this, most synthetic phosphate fertilizers advertise their product as water-soluble. What better way to get this vital nutrient where it needs to be – to the roots of an actively growing plant – than to suspend it in the irrigation water that the plant also uses? The problem is, that only allows for around half of the phosphorous applied to the soil to reach the plants that need it. In sandy or loose-textured soils, this phosphorous is prone to run off into adjacent water sources, resulting in eutrophication and other unsavory ecological effects.

In terms of ecosystem services, it makes sense that plants and animals would figure out over the last several billion years of evolutionary history how to access such a valuable nutrient. One main way that terrestrial plants access biologically fixed phosphorous in soils is by partnering with arbuscular mycorrhizal (AM) fungi. In fact, an estimated 80 percent of all plant species on Earth make use of this symbiotic fungal relationship.[6] The long, threadlike strands of AM fungal hyphae extend the reach of plant roots and allow for increased uptake of nutrients like phosphorous. In exchange, the AM fungi tap into plant roots for sugars. Some plants, such as the invasive spotted knapweed, form this beneficial mutualism quickly and efficiently.[7]

Knapweed loves disturbed places, and it thrives in abandoned agricultural fields, hay fields, and overgrazed pastures in the intermountain western United States. It also invades undisturbed grasslands, pointing to a potential successional role of the plant in landscapes where disturbance regimes have transformed from historic norms. The plants are efficient accumulators of soil-borne phosphorous and are thought to compete for it with both desired crop plants and native plants on rangelands. However, new research shows that knapweed and its mycorrhizal associates are so effective at accessing available phosphorous that they increase phosphorous availability at invaded sites.[8] This study showed that spotted knapweed contained twice as much phosphorous in its tissues compared to three species native to the same area. Though this research could be used to demonstrate the capacity of spotted knapweed to use up limited phosphorous resources (which other studies have done), the authors go on to show that sites invaded by spotted knapweed were also higher in soil-borne phosphorous. Though the knapweed has a greater ability to accumulate phosphorous, it is not at the expense of native plants or other surrounding vegetation, and its ability to accumulate the nutrient may even contribute to increased overall biological availability.

Spotted knapweed and its mycorrhizal associates are efficient scavengers and dispersers of soil-borne phosphorous and are invading systems whose design and implementation inhibit or degrade this critical ecological service. Knapweed also provides a number of other ecological services. Like many members of the Asteraceae family, knapweeds are a favorite of pollinators, providing nectar for many species of butterflies, flies, and bees – including native pollinators like bumblebees and digger bees – throughout their long blooming season.[9] Knapweeds are so popular with bees that the flowers represent a major source of honey production in the inland western United States. Finches, chipmunks, and deer mice, which in turn provide food for various species of owls, hawks, and other predators, eat their seeds. Other animals, such as California and Rocky Mountain bighorn sheep, white-tailed deer, mule deer, and elk have been reported feeding on diffuse and spotted knapweed seed heads in British Columbia in the winter and early spring. In fact, knapweed seed heads are the most common component of the California bighorn sheep's winter diet when all other forage is covered by deep snow.[10] Knapweed, whose taproot can grow 20 feet deep – survives in compacted, degraded, drought-ridden, and nutrient-poor environments.

Despite these valuable functions that knapweed serves, the plant is regularly treated with the herbicides clopyralid, picloram, and 2,4-D on rangelands and in the native grasslands where it is found. Especially with "peak phosphorous" looming in the relatively near future, we should consider how to make use of the ecological services provided by species like spotted knapweed. Though livestock tend to avoid mature plants, they are known to preferentially graze the young growth. One study found that intensively managed and well-timed pulses of grazing sheep reduced populations of knapweed with minimal effects on native balsamroot and Idaho fescue.[11] This practice would cycle the valuable phosphorous contained in the knapweeds' tissues back into the soil in the form of animal manure and urine. Over time, though knapweed will probably never be eliminated, its populations will be brought under control while achieving other valuable ecological yields.

Knapweed could also be used in buffer strip plantings in agricultural areas to intercept phosphorous-rich runoff before it reaches waterways. Its efficiency in taking up the excess phosphorous that leaks from the system would be invaluable. Beekeepers could make use of the nectar, and the plant material could be mowed and used as phosphorous-rich compost to add back into the fields. This simple step would return some portion of the valuable and rapidly depleting mined phosphorous resource. Even better would be to figure out unique polycultural associations that feature knapweed as part of the crop production system, as the efficient phosphorous-seeking fungal root extensions will pump this nutrient to surrounding plants and decrease the need for synthetic phosphorous fertilizers.

Looking deeper at the ecological services provided by invasive species has the potential to elucidate heretofore unrecognized relationships, as well as the potential for creative management approaches. Species like spotted knapweed are not present in ecosystems because of some unknown, random, and ecologically malignant purpose. Everything gardens, and knapweed is "gardening" phosphorous in the places where it is needed most. Learning to see, integrate, and make use of these features is a critical task for the future.

Pollinators

The ecosystem service of pollination is an intricate dance of plants and the insects, birds, and mammals (and occasionally wind and water) that transport and mix the plants' procreative materials. Some food crops, mostly the grasses like wheat, corn, rye, and oats, are wind pollinated and do not require insects to transfer their genetic material from plant to plant. At least one-third of all commonly eaten foods, from almonds to blueberries to cucumbers, rely on pollination to produce the edible portion of the plant. Other crops, like lettuce, kale, and carrots, do not require pollination to produce the parts of the plant we eat, but they do require pollination to generate seed year after year. Even foods that are not generally grown from seed, like potatoes, roses, and garlic, benefit from pollination to create new and improved varieties with better pest or disease resistance.

Bees and wasps, beetles, bugs, bats, birds, flies, gnats, mosquitoes, moths, butterflies, and even slugs and snails pollinate plants. Worldwide, approximately 200,000 organisms pollinate an estimated 250,000 to 400,000 species of flowering plants. By contrast, only 550 species of nonflowering plants, including grasses and conifers, depend on wind and water for pollen dispersal. The coevolution of plants and pollinators has led to the proliferation of global floral and faunal diversity. A flower's shape, size, color, smell, and bloom time attracts the type of pollinator most suited to carrying out the important task of genetic combining. Moths and bats pollinate night-blooming flowers like cactus and datura. Some flowers smell like carrion and are particularly attractive to flies. Bees, butterflies, and hummingbirds – depending largely on flower structure and accessibility of the nectar, usually pollinate bright colored and showy blossoms. Hummingbirds, whose long beaks and tongues can reach far into the base of a conical flower, often pollinate long, tubular flowers. Flowers with flatter tops, like yarrow, are well suited to butterflies because they provide a convenient landing pad from which to sip the nectar.

Many invasive species are well loved by pollinators. Himalayan blackberry, Japanese knotweed, kudzu, yellow star thistle, and others are known for their nectar-rich floral displays. As a result, these invasive species are beloved by beekeepers and honey producers. In many regions of the country, invasive species are now the only known source of pollen and nectar for native pollinators because of urbanization and the encroachment of agriculture into pollinator habitat. In California's Central Valley, twenty-nine of thirty-two native butterfly species breed on introduced plants and thirteen have no known native hosts in the region because of habitat destruction.[12] In many contexts, invasive species are the only available pollen resource.

Pollinators are in decline around the world, mostly because their importance wasn't given enough consideration as industrial agriculture and development boomed. Of course, this doesn't mean we don't need them. In central California, almond producers spend around $250 million to import honeybee hives to pollinate their orchards. This entails shipping thousands of hives from Oregon, Montana, and Idaho, setting them out in the almond orchards for three weeks, and then packing them back up and sending them on to the next nectar flow. Although this is a boon for beekeepers in the United States, it's stressful to the bees, leading some beekeepers to believe it makes the bees more vulnerable to pests and diseases like nosema, foulbrood, and colony collapse disorder. In attempts to manage these diseases, many beekeepers apply pesticides and fungicides to their hives.

One reason it's necessary to transport bees is because there's not enough year-round pollinator habitat near the almond orchards. However, some of the weeds in almond orchards – which have come to be known as invasive species in these production settings – include the geranium filaree, the native willow herb, sow thistle, fleabane, and mallows – all pollinator attracting species. These plants are treated throughout the year with preemergent herbicides like pendimethalin and postemergent applications of glufosinate and saflufenacil. Although these plants may not be enough to support a large population of honeybees throughout the year, their presence points to an elegant ecological solution for a seemingly intractable challenge.

Almond growers should take the ecological hint and develop continuous, year-round nectar resources for pollinators, enabling their farms to support a robust population of bees, eliminating transport costs, and diversifying potential income streams with sales of honey, pollen, and mead. The almond orchards themselves could be replanted to a diversity of tree crops, including honey locust, Asian pears, mulberries, persimmons, pistachios, and chestnuts that would diversify and extend the nectar season for native and on-site pollinators. Hedgerows of native and nonnative pollen- and nectar-rich plants interplanted among the crop trees would support the health of the orchard by providing habitat for beneficial insects, birds, amphibians, and reptiles, thereby reducing or eliminating the need for insecticide-based "pest control." Sheep, chickens, and pigs could graze rotationally beneath the trees, adding fertility as they turn weeds and spoiled fruit into edible protein. The weeds in almond orchards, and in all other agricultural operations, are a teaching tool. They challenge and encourage us as designers to develop truly ecological agriculture – agriculture that goes way beyond "ecological services" to a model so abundant that it is impossible to imagine all the yields.

Clean Water for Life

The planet's water cycle is an amazing process. Water continually moves from gas to liquid to solid as it evaporates, condenses, and freezes. Streams and rivers flow constantly to oceans, where water evaporates into clouds, falling as rain or snow and percolating through soil to flow once again into streams. Ninety-eight percent of the water on Earth is saline, held in the oceans where it is unusable to terrestrial life except when it is evaporated by surface winds and turned into rain.

In fact, only one-tenth of 1 percent of the water on the planet is available at a given time for life's necessities. In highly functional ecosystems, water filtration and purification occurs as rainfall slowly percolates through plant roots and organic matter–rich soils. The physical structure of the roots and the microorganisms living in the soil break down, transport, and uptake nutrients, sediments, and other contaminants that may be present in rainfall or runoff. In healthy stream channels, fast-moving waters ripple over stones adding oxygen for the aquatic organisms.

Most land use planning and design does not adequately preserve the functionality of natural filtration systems. Impermeable structures like roads, parking lots, and roofs direct runoff into storm drains, which add polluted water to streams and rivers. Sewage effluent treated with chlorine flows into lakes, creeks, and estuaries. Irreplaceable freshwater resources are treated as dumps, and the deterioration of habitat and overall ecosystem health are externalized. Most surface freshwater sources, including rivers, lakes, and streams throughout the world, are severely degraded to the point where they lack the ability to support life.

> The alarming trends for freshwater fauna are linked to extensive habitat deterioration caused by sediment loading and organic pollution from land-use activities, toxic contaminants from municipal and industrial sources, stream fragmentation and flow regulation by dams, channelization and dredging projects, and interactions with increasing numbers of exotic species. Of 5.2 million km of stream habitat in the contiguous United States, <2% (<100,000 km) is of sufficiently pristine quality to be federally protected, and only about 40 rivers >200km long remain free flowing. Such massive deterioration threatens some of the world's richest freshwater faunal assemblages.[13]

One of the world's greatest freshwater aquatic ecosystems is the Great Lakes, which straddle the border between the United States and Canada and contain one-fifth of the world's freshwater. The lakes are situated within a highly industrialized agricultural and manufacturing-based economy, and their water is used to manufacture fertilizers, plastics, and batteries, as well as in the production of coke tar, where it is used to wash ammonia from the coke oven gas. These manufacturing facilities, as well as power plants, take in and release water through buried pipes along the lakeshore. Populations of native fish, mollusks, and bivalves have declined precipitously over the past hundred years after the development of these industries. There have been 281 species of native mussels in the Great Lakes; nineteen are now extinct, twenty-one are thought to be extinct, seventy-seven are endangered, forty-three are threatened, and seventy-two are of special concern. And one is thriving, though it is not native.

The zebra mussel was one of the species that prompted the creation of the National Invasive Species Council because of its incredible proliferation. In 2004, the Lake Huron Center for Coastal Conservation found that a population of zebra mussels covering one square meter increased from one thousand to seven hundred thousand individuals in six months. Their sheer numbers and propensity to attach to any hard surface, including boats, buoys, docks, and the intake and pollutant-laden effluent pipes present challenges to industry and recreation. They also attach to the shells of native mussels and inhibit their ability to burrow, although no native mussels have gone extinct since the introduction of zebra mussels in the 1980s. The true biological consequences of zebra mussels on the Great Lakes ecosystem are still poorly understood.

Zebra mussels are filter feeders, and when they open their shells to admit water to extract plankton, they also take up detritus and sediment, and release clean water. A single dime-sized zebra mussel can filter up to one quart of water every day. Their filter feeding habits increase water clarity, resulting in further sunlight penetration and enhanced vegetative growth, including of native algae. Zebra mussels have been found to concentrate free-floating copper, nickel, and zinc from the water column and deposit the metals into lake-bottom sediments where they are less biologically available.[14] They remove carcinogenic and mutagenic persistent organic pollutants including polychlorinated biphenyls (PCBs), polycyclic aromatic hydrocarbons (PAHs), and dioxin from the water and store them in their tissues. Zebra mussels have also been found to accumulate *Giardia*, *Cryptosporidium*, *Salmonella*, Hepatitis A, and strains of antibiotic-resistant bacteria including *E. coli* and *Enterococcus* spp.[15] In all cases, concentrations of these viruses, bacteria, heavy metals, and organic pollutants were found to be greater within the mussels than in the surrounding water, meaning they are taking the substances out of the water and keeping them out.

Although the intake of heavy metals slows down the rate at which zebra mussels filter water, they are able to survive and even thrive in highly polluted bodies of water. Native mussels, by and large, can't do this. Of the 281 species of mussel that once thrived in US freshwater ecosystems, a total of 253 are extinct, endangered, or species of concern because of habitat and water-quality degradation due to the introduction of heavy metals, organic pollutants, dredging, increased sedimentation due to excessive erosion, and overharvesting for the historic button industry. These changes in aquatic ecosystems also decimated native fish populations, which native mussels use as substrates for their continued growth. Without the presence of large populations of fish, the reproductive capacity of native mussels declined, contributing to their near extinction. Zebra mussel larvae, on the other hand, float freely in the water until they find a hard surface to adhere to and complete their metamorphosis – they do not require fish to transport them. They are perfect candidates for the existing ecological conditions in the Great Lakes, and their filtration activities are enhancing freshwater food webs in the region.

Many native fish, birds, and other animals eat young and adult zebra mussels. Migratory ducks have changed their flight patterns in response to zebra mussel colonies.[16] Lake sturgeon feed heavily on zebra mussels, as do yellow perch, freshwater drum, catfish, and sunfish. The increase in aquatic plants due to increased water clarity provides excellent nursery areas for young fish and other animals, leading to increases in populations of northern pike and smallmouth bass in Lake St. Clair.[17] One of the major concerns with the mussels is that by filtering out the phytoplankton and making the water more clear, they decimate the food source of valuable sport fish like walleye and perch – but these populations have remained stable or increased since the mussel invasion.[18] The mussels are also used as indicators for the presence of toxins in the water since they can be easily tested for persistent organic compounds and heavy metals. One study estimated that the mass of zebra mussels in Saginaw Bay, Lake Huron, contained 1 metric ton of PCBs within their bodies, approximately equal to the total amount of PCBs found in surface sediments in the bay.[19] Though there is concern that the accumulation of toxins will spread throughout the food web and harm other species, this remains to be proven. A rigorous study undertaken by the EPA showed that the majority of the PCBs cycled through the mussels accumulated in the lake-bottom sediments where they were less available to higher trophic level feeders like fish and birds.[20]

The zebra mussel invasion brings up a number of interesting issues related to ecosystem services. Water filtration is generally considered a beneficial service rendered by organisms including bacteria, plants, fungi, and some animals like the mussels. It seems like this should be viewed as a good thing in highly polluted ecosystems such as those found in the Great Lakes region. Is it better to have free-floating persistent organic pollutants, heavy metals, and antibiotic-resistant bacteria readily making their way into the food web and municipal water supplies or to have them concentrated into organisms where they can be easily harvested, collected, and more intelligently disposed of?

Zebra mussels are already serving important ecological functions that could easily be enhanced by using them specifically as bioremediation agents. Inserting easily moveable hard substrates like pipes, rods, or planks in specific locations specifically for the mussels to colonize would give them a preferred habitat, potentially decreasing their colonization of native clams and mussels. Once their populations grow sufficiently large on these surfaces, they could be removed by scraping, and their bodies, along with the accumulated contaminants they hold, could be collected.

What to do with the accumulated toxins is a persistent issue in the field of bioremediation – one that will probably be a huge industry in the future. Perhaps we could compost the accumulated mussel shells in large, mycelium-laden piles where their constituent parts, including both toxins and valuable minerals like calcium, could be broken down for reuse or rendered inert. Paul Stamets, author of *Mycelium Running*, describes how mushrooms can be used to break down a wide variety of toxins, rendering them biologically inert or making them easy to recapture and recycle. *Xylaria* spp. have been found to neutralize strains of antibiotic-resistant bacteria; they and other species with similar characteristics could be used to turn these potentially lethal ingredients into beneficial organic material.[21] Oyster mushrooms can break down PCBs and other persistent organic pollutants.[22] Many common mushrooms, including the king bolete and the little button mushroom, can absorb heavy metals like lead, copper, cadmium, and mercury.[23] If these or other mushrooms were used as part of a mycoremediation strategy, their fruiting bodies where the metals accumulate could be dried, incinerated, recaptured, and recycled – and hopefully not returned to aquatic and terrestrial ecosystems again.

In order for this to work, the compost piles would require the addition of more carbon and nitrogen to ensure good decomposition. These could include sawdust from sawmills, fish waste from fish-processing plants, spent grain from breweries, and residential collection of kitchen scraps. All of these ingredients are readily available in the Great Lakes region. The intentional harvesting of pollutant-laden mussels could be the first step in the long process of restoring the incredible natural resource of the Great Lakes ecosystem. Poisoning them with copper or chlorine, which is a standard practice in the region, is not going to solve any of the underlying ecological issues that they highlight for us so brilliantly.

Watershed Restoration

Aquatic ecosystems have been largely mismanaged, and the growth of invasive species demonstrates the lack of appropriate interaction with these valuable natural resources and the ecological services they provide. Invasive species proliferate in polluted, nutrient-enhanced, or otherwise compromised waterways. They also exist in ecosystems that appear somewhat intact, though historic analysis can tell a different story and helps to explain the presence of some of the more vexing invasive species, such as Japanese knotweed and giant reed.

Beavers shaped the watersheds of North America over the course of the millions of years they have been present on the continent. These busy ecosystem engineers altered stream flow and shaped aquatic ecosystems from high mountain streams to alluvial floodplains in the process of building their homes. Beaver dams are unique structures – semipermeable, strong enough to walk on, and often inhabited for generations. They build these structures from trees, sticks, and mud, stacking them together so that the bulk of the stream flow is held behind the structures, but also with spillways and passageways so that water can move through without damaging their homes. The inside remains dry, though the entrance is submerged. These physical characteristics slow down floodwaters and capture sediments.

Prior to the arrival of European colonists, beavers inhabited creeks and wetlands from the arctic tundra of Alaska and northern Canada to the deserts of northern Mexico. Beaver trapping was one of the first industries to take hold on the newly "discovered" American continent—in many cases, fur trappers were the first people of European descent to explore the land west of the Mississippi and north of the Hudson Rivers. Their hunting and trapping, which fed the demand for beaver coats and felt hats in Europe, led to the near extirpation of the species by 1900.

Hunting and trapping of the beaver in seventeenth- through twentieth-century North America had extraordinary effects on the ecology of the continent, from the physical structure of riparian areas and floodplains to the species composition and richness of aquatic ecosystems.

> It is estimated that as many as two hundred million beavers once lived in the continental United States, their dams making meadows out of forests, their wetlands slowly capturing silt. The result of the beaver's engineering was a remarkably uniform buildup of organic material in the valleys, a checkerboard of meadows through the woodlands, and a great deal of edge, that fruitful zone where natural communities meet. Beavers are a keystone species, for where beavers build dams, the wetlands spread out behind them, providing home and food for dozens of species, from migrating ducks to moose, from fish to frogs to great blue herons.[24]

Where beaver populations are undisturbed, they have been observed to construct around ten (but up to seventy-four!) dams per kilometer of stream.[25] Where water once moved slowly and placidly through a series of beaver-constructed pools as it made its way from ridge top to river mouth, in the absence of these permeable dams, water moves faster through the channel, without as much time to soak in for use by surrounding vegetation. Beaver dams also influence off-channel hydrology by making the environment more moist. A beaver dam in Minnesota measuring 6 feet by 250 feet influenced the hydrology within half a mile of its location. Lack of beavers means fewer semipermeable dams, less water volume soaking into the surrounding land, and, therefore, reduced aquifer recharge.

It is hard to imagine a time when streams and rivers pulsed and changed with the natural patterns of flood and drought, their courses shifting and meandering in braided networks of side channels, rather than confined to a singular path. In the absence of beavers, and concomitant with the construction of impermeable concrete dams and other significant alterations in hydrology, streams and rivers are now constrained to a specific space, where their flows continually erode sediments, unmitigated by side channels, back eddies, or flow-reducing structures like beaver dams. Today, beavers are considered nuisance pests in many areas because their dam building and resultant flooding damages property.

A lack of beaver dams or similar structures in a waterway does not mean there is less water moving through the channel, but that the same amount of water moves downhill with greater force, and the sediment that the water carries moves quickly to the ocean. Rapidly-moving water carrying sediment is often turbid, as the sediment is not collected by any flow-decreasing structure. The increased velocity of water also contributes to the channelization and scouring of stream banks, many of which today are highly aggraded with steep channels. Beaver-structured creeks and their clear, slower moving waters yield perfect spawning grounds for salmon and other anadromous fish, like trout, char, grayling, and whitefish, which require clear, cold, relatively slow-moving water with gravels of a certain size to lay their eggs. At least eighty species of fish, including many that are critically endangered like sculpins, sticklebacks, and suckers, have been found in beaver ponds, where they benefit from the slow-moving water; they don't have to expend so much energy fighting against the currents in search of food.[26]

Beavers were also instrumental in creating some of the best agricultural soils on the continent – floodplain soils. Creek and riverside farms may owe their fine fertile soil to beavers, since water slowing down and spreading out behind a beaver dam widens the lens of water and the sediment it carries to beyond the stream channel. Over time, the depositional activity results in a series of flat terraces moving down the stream channel in a stair step pattern. One study found that beaver dams retain from 35 to 6,500 cubic meters (m^3) of sediment behind their structures.[27] If this material were used to fill 55-gallon drums, 32,500 would be required to capture the sediment behind the largest beaver dams. Over time, streams turn to wetlands, and wetlands to meadows, meadows to prairies, and eventually, a long time after the beavers are gone, to farms.

Beaver dams are considered ecologically beneficial because they conserve water, nutrients, and sediment and especially because the dams hold water higher up in a watershed before it makes its inevitable way downstream. The secondary effects of these ecological services promote the ability of other species to thrive, including fish, insects, birds, mammals, and a wide variety of riparian vegetation that depend on the riparian channel profile created by busy beavers. The total sum of ecosystem services yielded by beavers is difficult to quantify, but one thing is certain – the unprecedented reduction in their populations through historic hunting and trapping has rippled throughout the riparian ecosystems where they were once found and contributes to the modern proliferation of riparian invasive species.

Interestingly, many of the same traits are assigned to the invasive giant reed but are viewed as negative ecological effects. Giant reed is targeted as a noxious invasive weed in California waterways because it increases local flooding by spreading water laterally beyond a channel, increases sediment retention, and changes the direction and force of in-stream currents and therefore decreases channel velocity.[28]

In the absence of an animal whose engineering feats increased sediment retention, spread water beyond the channel, and decreased water flow velocity, there is a plant growing that does many of the same things. As Parmenides mentioned nearly two thousand years ago, nature abhors a vacuum. Robust stands of giant reed do preclude the growth of native riparian vegetation. They also don't provide habitat for the diversity of animals that native species do, but given the alteration of stream structure engendered by the loss of beavers and other related changes in land use and hydrology, there is no indication that native riparian vegetation, including flood-adapted willow and cottonwood, would survive in the altered conditions.

Short of widespread beaver reintroduction, which should also be a serious consideration for restoration of North American waterways, making use of the physical and structural characteristics of invasive riparian species like giant reed and Japanese knotweed is a viable option that over time, could produce similar ecological results. Giant reed is one of the most prolifically growing plants on the planet. It can grow up to 7 centimeters per day and produces 60 tons of biomass per acre. Its canes are similar in strength to bamboo and, when cut, would make suitable material for building woven weirs across stream channels. Riparian restoration often calls for the placement of gabions – semipermeable rock-filled cages behind which water and sediment can pool – in streams. Gabions are extremely effective for enhancing stream structure and flow, but they are energy and resource intensive. Rocks are often trucked in to the site at great expense, and eventually the metal cages that hold them will rust away, leaving huge piles of rocks in the reemerging stream corridor. A more ecologically based approach would make use of giant reed's abundant biomass by following the permaculture principle of using biological resources.

Sturdy anchor canes can be placed directly into the streambeds to provide the initial structure, much like the ribs of a basket. Additional canes can be woven between these, creating a semipermeable, long-lasting (but still decomposable) structure that encourages the deposition of sediment and pooling of water. These structures can be made similar in shape and size to beaver dams, around 6 feet tall and 5 feet deep, and as wide as the span across the stream channel. Once the woven structure is made, the prolific leafy biomass of the reed can be layered into the basket. Over time, the material will break down, and the rhizomatous reeds will resprout. At this point, the canes can be cut again, adding further material to the water retention structure. If these natural gabions were placed at the same concentration as historic beaver dams (10–70 per kilometer), not only would the population of giant reed decline, but the stream would also start to change in character and ecological quality.

By consistently pruning back its photosynthetic leaf surface, over time, the vigor of the reed's rhizome would decrease. The decomposing reed mulch in the gabion would turn to soil over time and could eventually be planted with willows and cottonwoods, providing shade and habitat to the riparian corridor. The gabion structures, like beaver dams, allow water to pass through, but build up sediment behind them. Eventually, the stream channel would attain a stair step gradient again, the channel would widen, and more water would stay in the watershed longer. Seasonal floodplains would form, and soil would be kept from the sea. Native riparian vegetation would flourish in the rehabilitated stream channels.

Turning Problems into Solutions

Invasive species provide opportunities to study the dynamics of ecosystem change given serious declines in ecological function engendered by modern land use practices. Nitrogen-fixing invasive species are viewed as ecologically destructive, when in fact nitrogen is one of the most valuable plant growth nutrients. Rampant growth of invasive species is seen as threatening to ecosystems, when turning diverse ecosystems into monoculture crops destined for human or animal consumption is considered normal. Pollen- and nectar-bearing plants are considered weeds in agricultural operations that spend millions of dollars every year to ensure pollination. Zebra mussels are seen as threatening to the aquatic ecosystems of the Great Lakes, though the presence of PCBs, dioxin, heavy metals, and antibiotic-resistant bacteria in the water should be of much greater concern and galvanize at least as much fervor and activism as the incursion of inch-long mollusks.

There is something wrong with this picture, and the time has come to acknowledge the ecologically beneficial role that invasive species play in the presence of myriad life-degrading processes and in the absence of well-designed land stewardship models. Species invasions don't just happen to passive, unchanging ecosystems; in some cases, they are the direct result of fluctuations in the larger environment, systemic responses to imbalance or crisis. Invasive species represent a call to engage in creative management that takes these ecosystem services into account. If there is a strong desire to remove or slow the spread of an invasive species, then the ecosystem services that they fulfill must be taken into account as part of the management plan.

An objective view of invasive species sees them in terms of patterns, processes, and functions that they represent, and it is important to acknowledge that they are not necessarily "good," just as they are not necessarily "bad." A whole systems perspective on invasion must necessarily move beyond the dualism imposed by human preferences for certain ecological outcomes. This does not necessarily imply amnesty for them, but rather provides a framework for long-term ecologically rational management of their populations that addresses the processes contributing to invasion.

CHAPTER 6

Everyone Gardens

"We are all indigenous to this planet, this garden we are being called on by nature and history to reinhabit in good spirit. To restore the land one must live and work in a place. The place will welcome whomever approaches it with respect and attention. To work in a place is to bond to a place: people who work together in a place become a community, in time, grows a culture. To restore the wild is to restore culture."

— *Gary Snyder*

When Fr. Bartolomé de Las Casas arrived on the island of Hispaniola in 1502, he noticed the native Taino people making paper and cloth from mulberry bark and remarked that there were "as many mulberries as weeds" growing on the island.[1] Mulberries are native to China, where people have made textile products from the multifunctional trees for thousands of years. Las Casas noted their prolific growth and widespread use a mere ten years after Columbus made his first incursion into the Americas, decades before their planned introduction for the same purpose. Other research shows that mulberries were well known and heavily utilized in relatively "uncontacted" tribal areas of Mexico and that the tools and techniques used to make paper from their bark were identical to those found in Southeast Asia.[2] The Maori people carried mulberry starts with them when they left Polynesia to settle New Zealand more than 1,200 years ago and planted them to continue their papermaking tradition. Mulberries were introduced and incorporated both culturally and ecologically into the Americas and other lands well before Columbus and his contemporaries weighed anchor to mark the beginning of the most recent period of colonization.

Though Columbus and other explorers from the late fifteenth and early sixteenth centuries are typically credited as the first people technologically capable of transoceanic seafaring, expeditions funded by European monarchs followed a long history of people crossing oceans in search of new lands. People traveled to Australia from Papua New Guinea or Timor 50,000 years ago.[3] And according to some research, there is evidence from 840,000 years ago for *Homo erectus*, the progenitor of our species, demonstrating "almost habitual use of navigation" between islands in Southeast Asia.[4] Bill Mollison notes that Polynesians used pattern maps that "lacked scale, cartographic details, and trigonometric measures, but nevertheless sufficed to find 200–2,000 island specks in the vastness of the Pacific Ocean! Such maps are linked to star sets and ocean current and indicate wave interference patterns; they are made of sticks, flexed strips, cowries and song cycles."[5] Though these ancient travelers didn't use written language, they employed sophisticated understanding and application of natural patterns.

People transformed the ecosystems they encountered as a matter of course, not least through the introduction of culturally important species of plants and animals they brought along with them on their migrations. Seeds, tubers, and cuttings of particularly hardy, delicious, or nutritious species that people knew how to cultivate and use made the journey with these early explorers. A text from Easter Island refers to "path of the sweet potato" from the west, and "path of the breadfruit tree" from the east, a botanically accurate assessment indicating the directions from which these plants came to the island. Breadfruit is native to Indonesia and Polynesia, and sweet potatoes likely originated on the west coast of South America. That these botanical pathways intersect on isolated Easter Island, 2,300 miles from the coast of Chile and 2,600 miles from Tahiti, demonstrates that pre-Columbian seafarers successfully traveled vast oceanic distances. Recent DNA analyses have confirmed the origins and transoceanic trade patterns of these culturally important food crops.[6]

In their book *World Trade and Biological Exchange before 1492*, authors John Sorenson and Carl Johannessen provide evidence for at least a hundred plant, animal, bacteria, and fungal species that were traded across continents and oceans – between Asia, Africa, and the Americans – for thousands of years before the beginning of the modern colonial era. Among the plant species for which there is definitive evidence of precolonial transport between continents are corn, cotton, amaranth, agave, cashew, pineapple, milkweed, chili peppers, papaya, coconut, squash, turmeric, sweet potato, marijuana, sunflower, mulberry, basil, kidney beans, guava, sugar cane, and potatoes.[7] Animals and other fauna include dogs, mealworms, soft-shelled clams, and chickens.[8] A host of bacteria and diseases were likely part of these exchanges, as well as insects and other "pest" organisms hitching rides on their host plants and animals.

And paleobotanists are still attempting to understand, for example, whether tarweed is native to either Chile or Oregon. Indigenous people in both Chile and Oregon use tarweed as an important oilseed staple crop, and it grows as a native "wildflower" in both regions.[9] It is unlikely that the same plant species originated independently in both locations, suggesting that at some point, the seeds of this valuable staple food were exchanged and made their way up or down the coastlines of North and South America. The evolutionary origins of plants and animals are important, but knowing that people have been making contact with each other, sharing valuable species, and transforming their environment for millennia summons even deeper ecological and anthropological understanding and appreciation.

Although ecosystems around the world were shaped by species introductions on an unknown scale prior to European conquest, this moment of "contact" remains the standard by which an organism is judged native or nonnative. Native species are considered those that have existed naturally, free of the type of human-driven disturbances that commenced with European colonization. A native seed company in Missouri defines a native plant as "one that occurs naturally in a particular region, state, ecosystem, and habitat without direct or indirect human actions. Its presence and evolution in an area are determined by climate, soil, and biotic factors."[10] Unfortunately, definitions like this fail to account for the fact that the Americas and other lands that were "discovered" by European colonists were highly modified and managed by humans before that time through the intentional selection, cultivation, and propagation of the plants and animals that are today considered native.

"Nature's" Bountiful Garden

In many places, pre-Columbian cultivation and ecosystem management and modification occurred on a scale that affected the trajectory of evolution. According to ethnobotanist Nancy Turner,

> These plant management principles and practices represent a continuum of people-plant interactions, with differing levels of intensity and differing impacts on landscape. At the foraging end of the scale, plant-harvesting activities had little or no impact on the plant populations or the landscape. In contrast, activities associated with plant cultivation and domestication structured the species composition and diversity and thus had impacts at the population, community, and landscape levels, altering the natural (that is, successional) state, and creating anthropogenic landscapes.[11]

Ecosystems from Southeast Asia to Africa to South America were influenced by cultivation activities for thousands of years. Throughout the tropics, there is evidence of *terra preta* or black earth, a human-manufactured soil enriched by charcoal and fired clay fragments. Some of these are over seven thousand years old, though it is unclear whether these oldest samples were intentionally produced or by-products of human habitation.[12] Terra preta soils retain nutrients and help build soil organic matter and are as valuable today as they were thousands of years ago. Even vast swaths of the Amazon rainforest are dominated by culturally valuable species like babassu palm, lianas, Brazil nuts, and bamboo species – which were (and still are) planted and propagated by people.[13]

Some of the world's most diverse ecosystems are places that are inhabited by humans. Though some of their practices have changed in modern times, the Amazon basin is home to hundreds of tribes that practiced burning, selective cutting, planting, and weeding in order to create patches of mostly perennial gardens, which flow seamlessly into what is commonly (and erroneously) considered an undisturbed or "natural" land.[14] Anthropologist William Denevan studied the cultivation strategies of the Bora tribe in the Peruvian Amazon and found that their gardening style, which shifted garden patches through time and space, mimicked natural succession using important edible, medicinal, and material-bearing crops. Though they practiced swidden agriculture (formerly known as "slash and burn"), the results of this method over time actually produced highly diverse and productive perennial ecosystems featuring staples like cowpeas, manioc, pineapple, and peanuts grown among peach palm and guava. Eventually, large, long-lived trees like breadfruit and umari dominated the canopy, shading out the annuals and short-lived perennials. By this time, another patch nearby had been cleared for these shorter-lived crops. Although the site was initially burned, after twenty years, the site was home to at least twenty-two tree species yielding edible fruit, medicine, construction wood, and other materials that could be harvested for the next twenty to thirty years before the cycle started anew.[15]

Similar approaches to land management have been practiced for millennia by indigenous people in Africa, Indonesia, Asia, Australia, Polynesia, and the Americas, as well as in Europe to some degree. For example, the Yucatec Maya of Mexico selectively cut the forests where they live and manage the successional trajectory of the postdisturbance ecosystem over a fifty year cycle.[16] Some agroforestry systems in Indonesia and northern India follow similar patterns.[17] In this way, people achieve yields of different products over time from a single site, and the region as a whole features a mosaic of different successional stages.

Such activities use planned moderate disturbances to produce yields of useful products. A patchwork of early, mid-, and late-successional ecosystems also yields high biodiversity since it creates the greatest possible number of niches, which support the greatest diversity of plants and animals. Species that prefer shade will live in the older ecosystems, while those that prefer high-light environments thrive in the recently cleared patches.

The scale and degree of land modification practiced by precolonial indigenous cultures is still not widely acknowledged, although there is ample ecological and anecdotal evidence. Some assume that indigenous people are and have always been passive ecological actors – that their lifestyles have, and have had, minimal influence on their homelands. For example, a 1991 Smithsonian Institute book asserts "pre-Columbian America was still the First Eden, a pristine natural kingdom. The native people were transparent in the landscape, living as natural elements of the ecosphere. Their world, the New World of Columbus, was a world of barely perceptible human disturbance."[18]

The belief that the biological resource wealth European explorers and colonists encountered on their arrival to the Americas – the lush prairies, verdant forests, abundant fish and game – was the result of untouched nature continues to pervade modern policy agendas. And the idea that people should be kept separate from nature in order to preserve it was enshrined by the Wilderness Act of 1964: ". . . a wilderness, in contrast with those areas where man and his own works dominate the landscape, is hereby recognized as an area where the earth and its community of life are untrammeled by man, where man himself is a visitor who does not remain."

Some of America's most celebrated poets, authors, and philosophers champion the idea of wilderness and indeed the setting aside of nature for its own sake – free of human use except for tourism – is a unique feature of American land policy. John Muir famously wrote that "thousands of tired, nerve-shaken, over-civilized people are beginning to find out going to the mountains is going home; that wilderness is a necessity."[19] But the unblighted "nature" that so entranced Muir and others is mostly a myth. Most land was inhabited and tended by people, its abundance and beauty a product of careful attention. In fact, some researchers believe that every ecosystem in North America was shaped to some extent by human modification, which presents a much different picture of nature than that commonly described in most modern land conservation circles.[20]

John Muir was raised on a farm in Wisconsin, where he watched poor agricultural practices degrade the land to the point where his family had to move on to another piece of productive land. As an adult, he traveled to Canada, Florida, New York, and made his way to California via the Panama Canal, arriving in the Yosemite Valley in 1867. It was there that his vision of nature unobstructed by human interaction crystallized. He described the valley floor's lush floral diversity "as if Nature, like an enthusiastic gardener, could not resist the temptation to plant flowers everywhere."[21] He noted that

> the vegetation is exceedingly rich in flowers, some of the lilies and larkspurs being from eight to ten feet high. And on the upper meadows there are miles of blue gentians and daisies, white and blue violets; and great breadths of rosy purple heathworts covering rocky moraines with a marvelous abundance of bloom, enlivened by humming-birds, butterflies and a host of other insects as beautiful as flowers. In the lower and middle regions, also, many of the most extensive beds of bloom are in great part made by shrubs – adenostoma, manzanita, ceanothus, chambatia, cherry, rose rubus, spiraea, shad, laurel, azalea, honeysuckle, calycanthus, ribes, philadelphus, and many others, the

> sunny spaces about them bright and fragrant with mints, lupines, geraniums, lilies, daisies, goldenrods, castilleias, gilias, pentstemons, etc.[22]

What Muir didn't understand was that the Yosemite Valley had been home to the Ahwaneechee people, a branch of the southern Sierra Miwok culture, for thousands of years prior to their forcible removal in 1851, sixteen years before he happened upon a field full of flowers. The awe-inspiring landscape Muir described was not the work of some mysteriously gardening "Nature," but rather the result of consistent, thoughtful, and protracted stewardship by generations of people inhabiting an ecosystem full to the brim with edible, medicinal, and otherwise useful plants.

This story is unfortunately similar to others from throughout the Americas. Still, the concept of untouched and uninhabited wilderness persists. Though much of the surrounding land appeared unpopulated and "wild," the forests, meadows, and woodlands had actually been shaped by human activities, including hunting, burning, planting, pruning, seed saving, and plant selection, over the course of millennia. These actions changed the ecosystems – the scale of which varied widely – from exerting pressure on birds and mammals through hunting to intensive land clearing for cultivation of annual and perennial crops.

Disease and the Rewilding of America

One of the biggest reasons people think the Americas represented an untouched wilderness is because so many indigenous populations were lost to disease. The decimation of people, and the subsequent loss of ecosystem management, led to major ecological changes well before the majority of European immigrants arrived on the continent. Epidemiological data and genetic evidence show that indigenous populations endured significant reductions in population after colonization, characterized by waves of diseases such as smallpox, bubonic plague, typhus, cholera, tuberculosis, and malaria, as well as associated stresses from wars, displacement, and malnutrition.[23] In addition to the unimaginable personal, political, and social turmoil this must have caused to indigenous communities, it also changed the intensity of land management.

Columbus, Cabrillo, de Soto, Cortes, Pizarro, and Coronado traveled throughout Mexico, South America, the Caribbean, and the southern

United States from 1492 until 1542, bringing pigs and cattle with them, and often releasing them to feed on the bounteous pastures and forests. European explorers had immune systems that had adapted for thousands of years to European agriculture, livestock, and the diseases they carry. These diseases spread rapidly through indigenous families and communities – many of whom never even saw a European colonist – via trade and travel long before "settlers" arrived to claim land as their own.

By the time greater numbers of immigrants began to arrive from Europe in the late seventeenth and early eighteenth century, the population of native people, in North America at least, appeared sparse. But current research suggests that between forty million and one hundred million people inhabited the Americas prior to the excursions of Columbus and other explorers.[24] Fr. Bartolomé de Las Casas wrote in 1502 that "All that has been discovered . . . is full of people, like a hive of bees, so that it seems as though God had placed all, or the greater part of the entire human race in these countries."[25]

Cities with populations rivaling Paris and London flourished throughout North, Central, and South America, including Cuzco, Tenochtitlan, and Tikal. In North America, the city of Cahokia, Illinois, is thought to have sustained a population of fifteen thousand people and potentially up to forty thousand at its peak in the year 1250.[26] Populations of this size would surely have had an effect on their surroundings as people sought to meet their needs for food, clothing, shelter, fuel, medicine, household goods, crafts, and religious items.

Chroniclers of the de Soto expedition commented on the numerous towns, bustling villages, and extensive agricultural fields as they passed through what would become Florida, Arkansas, Louisiana, Georgia, Mississippi, and Texas in 1539–1543. When the de La Salle expedition passed through the same area in 1682, just over 150 years later, they found sparse populations and abandoned fields.[27] By the time writers like Muir and Thoreau were expounding on the importance of an intact wilderness in the late nineteenth century, the ecology of the Americas must have appeared wild and untouched – because it had, by then, been unmanaged for hundreds of years.

Tree ring analyses on Douglas fir trees in the Willamette Valley of Oregon show that the oldest trees germinated around 1500, around the same time that first wave of smallpox probably made its way through the region.[28] When the land was managed as an oak savannah, rich in edible fruits, nuts, bulbs, and tubers, Douglas fir was considered a weed whose encroachment was kept at bay by consistent burning. Widespread incursion of Douglas fir dating from this time period indicates a significant shift in management. According to this same tree ring research, further proliferation of conifer trees followed subsequent waves of disease in 1770, 1826, and 1852. As the management changed, so did the ecology.

Indigenous people held the successional drive of the Willamette Valley ecosystem in check through consistent, moderate disturbance of low-intensity burns that favored the proliferation of the oak savanna. The oak savanna polyculture is an early to mid-successional system, with yields of abundant grass and forb forage for deer and elk, berries, acorns, camas, and harvest lily. Without constant, purposive intervention, the ecosystem tends toward a later successional conifer-dominated forest ecosystem. Today, forest-based plant communities are more common in the valley, while the oak savanna ecosystem is the most endangered plant community in the West.

This change in ecosystems due to population declines is not solely a West Coast phenomenon; fire management was practiced throughout the Americas as a cultivation tool in perennial systems. In eastern US forests, decline in fire management led to the proliferation of fire-sensitive tree species like red maple, sugar maple, beech, and black cherry, which created cooler, moister forest interiors and altered the species composition of the understory.[29]

Meat from deer, elk, buffalo, turkeys, rabbits, and other animals was central to the diets of temperate North American indigenous people. With a decrease in hunting, animal populations increased, and their feeding habits changed the ecosystems where they lived. For example, the de Soto expedition traveled for four years throughout the southeastern United States and did not record the sighting of a single buffalo.[30] Two hundred years later, buffalo were known to roam throughout the same area.[31] As the buffalo increased in population and expanded their range, their foraging habits encouraged the proliferation of open, prairielike ecosystems that would have grown into forest if left ungrazed. Large populations of ungulates also influence succession by encouraging the plant species that best suited their dietary need through grazing, wallowing, and deposition of urine and manure.[32]

Buffalo were hunted nearly to extinction in postcolonial America, which changed the successional trajectory of the ecosystems where they lived yet again. Newly arrived colonists also hunted populations of egrets, geese, mountain sheep, antelope, beavers, otters, and many other native animals, which altered ecosystem physical structure and food web relationships. Farmers, ranchers, and homesteaders also focused on the removal of wolves, cougars, bears, and other top predators, which allowed their prey to proliferate – resulting in still more changes in ecosystem structure. Early introductions of pigs, sheep, horses, and cattle were largely unmanaged and overstocked, allowed to graze from the timbered ridges of the Sierra Nevada and Cascade Mountains to the fertile valleys of the Mississippi and Ohio Rivers, leading to unprecedented and poorly understood changes in plant community assemblages.

The Myth of Wilderness

America's most celebrated wilderness areas were once peoples' homes, and many of the most prized native plants are remnants from gardens and orchards. Native plant lists from every state or region in the United States show an abundance of perennial edibles, medicinals, fuel and fiber plants, as well as pollen and nectar sources. Georgia's native plant inventory includes American chestnut, Allegheny chinquapin, hackberry, five species of hickory, hawthorn, hornbeam, honey locust, saw palmetto, red mulberry, eighteen species of oak, pawpaw, persimmon, sassafras, serviceberry, wintergreen, redbud, sourwood, and black walnut.

In Michigan, the list includes milkweed, strawberry, lobelia, bergamot, nodding onion, boneset, wild ginger, and blue cohosh, grown among linden, elderberry, black cherry, sugar maple, nannyberry, and highbush cranberry.

In Arizona, natives include cat's claw, agave, chia, mesquite, ephedra, ocotillo, hackberry, perennial chiles, barberry, palo verde, prickly pear, pinyon pine, lemonade berry, jojoba, wolfberry, and yucca throughout the landscape.

This proliferation of useful plants throughout the Americas was not a matter of chance but of purposive guidance by the people who lived in the regions for millennia. These plants have come to be known as native, wild, and natural, but in fact, they were intentionally cultivated. Like any garden, these species were carefully chosen, maintained, and propagated over generations until eventually the entire landscape was full of useful plants.

European colonists, who were accustomed to fields of annual grains and pulses along with domestic animals, seemed to have had no idea what they were looking at when they encountered these diverse perennial landscapes.

Acknowledging that people were, to a greater or lesser degree, responsible for the bounty and diversity requires a massive shift in our ideas of wilderness and nature. Instead of imagining no human presence on wild land, we might begin to see the patterns of their interaction and gardening. This awareness is important to how we understand changes in the ecosystem structure, including invasive species. Instead of viewing invasive species as disturbing pristine natural landscapes, we have to better understand how changes in historic land stewardship practices have also altered them. As Kat Anderson, ethnoecologist for the Natural Resource Conservation Service, relates in her book *Tending the Wild*,

> Interestingly, contemporary Indians often use the word *wilderness* as a negative label for land that has not been taken care of by humans for a long time, for example, where dense understory shrubbery or thickets of young trees block visibility and movement. A common sentiment among California Indians is that a hands-off approach to nature has promoted feral landscapes that are inhospitable to life. "The white man sure ruined this country," said James Rust, a Southern Sierra Miwok elder. "It's turned back to wilderness." California Indians believe that when humans are gone from an area long enough, they lose the practical knowledge about correct interactions, and the plants and animals retreat spiritually from the earth or hide from humans. When intimate interaction ceases, the continuity of knowledge, passed down through generations, is broken, and land becomes "wilderness."[33]

The idea that people can and should have a hand in the active stewardship of land in order to enhance its diversity and abundance is largely absent from conversations about how best to preserve endangered species and ecosystems. In fact, the opposite approach is often rigidly enforced: People and their activities are confined to areas not designated wilderness and wilderness areas grow less diverse and more vulnerable to invasion and catastrophes. As ecologist Fikret Berkes notes:

> Our conventional conservation blueprint excludes disturbance, and aims for unperturbed, stable systems in a state of equilibrium. 'Balance of nature' and equilibrium thinking support the view among some conservationists that the best way to conserve nature is to seek out high biodiversity, supposedly pristine ecosystems, remove all human influences (such as haying, grazing, collection of non-timber forest products, use of fire) and re-establish natural biodiversity by stabilizing ecological processes. Such an approach largely failed. Ecosystems are dynamic and disturbance, including some level of human use and human disturbance, has an important role in maintaining ecological processes.[34]

While the concept of wilderness has led to the protection of certain places from the resource extractive activities of industrial capitalism, it has also separated people from it in order to preserve the integrity and beauty of the protected land. In the current context, wilderness areas are vital repositories of plants and animals that would not survive conventional forestry and agriculture. They have become refugia for imperiled species and ecological communities that would have otherwise been turned under the plow, buried by asphalt, or pushed out of their territory by encroaching development.

Reinterpreting wilderness is critical to an objective understanding of invasive species, since it shows how peoples' purposeful management over time has shaped their nature and how the lack of consistent stewardship creates opportunities for invasive species to thrive. The US Forest Service states that "non-native, invasive species of plants and animals are invading, displacing and destroying native species in wilderness all across the country," a statement that treats ecosystems as passive objects under threat by outside forces. But ecosystems are dynamic associations of plants, bacteria, fungi, water, nutrients, soil, climate, and animals – including people – that live, die, and affect one another as these processes play out over time.

Invasions in Neglected Gardens

Low-intensity burning and other methods of indigenous management worked with ecological succession to encourage maximum diversity. A lack of periodic pulses of moderate disturbance leads to declines in biological diversity, which in turn makes ecosystems more prone to invasion.[35] Today, yellow star thistle, a maligned invasive species known for the sharp thorns that emanate from the flower head, is present along roadsides and in lower elevation grasslands in Yosemite Valley. Star thistle is often found in once-disturbed areas like overgrazed or abandoned fields and is of particular concern in rangelands, where it outcompetes more favorable fodder plants. Cattle are injured when foraging on its spiny growth, and horses can be poisoned after eating large amounts of the plant. The garden that Yosemite once was has been abandoned, creating niches for novel species like star thistle that thrive in neglected areas. In fact, research suggests that star thistle is particularly successful when invading California grassland ecosystems whose ecology has been altered by reduced fire frequency.[36]

This annual plant, like many in the Asteraceae family, produces numerous seeds and forms dense mats that inhibit the ability of other plants to germinate. It has a deep taproot, which provides the plant with significant drought resistance. Yellow star thistle nectar is loved by native pollinators as well as honeybees, which use it to make a light flavored honey. Several species of sparrows, towhees, goldfinches, California quail, and the pine siskin relish the seeds of this thistle as part of their diet. Yellow star thistle is native to the Mediterranean region of southern Europe, where it grows in grassland environments similar to those in California, Oregon, and Washington.

Star thistle grows in disturbed or compacted soils caused by overgrazing, road building, and other soil-compromising activities. Its deep taproot carries water and nutrients 3 feet below the soil surface, slowly improving the damaged soil. When the plant dies every year, the root decomposes, leaving a channel for water to travel down and a spike of organic matter reaching below the soil surface. Star thistle serves multiple ecological functions, such as soil stabilization, nectar and seed provision, and protection from uncontrolled grazing. Though it presents challenges to land uses such as grazing, star thistle is playing an important role in the ecologies where it is found, especially considering the damage to soil structure and diversity resulting from unmanaged grazing and lack of fire-based management.

Star thistle is especially ubiquitous in grasslands that were historically managed by fire, specifically the frequent, low-intensity burning practiced by indigenous people in these areas. Grasslands that are no longer managed by fire have lower plant diversity, and those that have been overgrazed have even less.[37] Careful reintroduction of managed fires at a test site in northern California showed a 90 percent reduction in star thistle populations, a 99 percent reduction of star thistle seed in the soil, and increased diversity and richness of native species over the course of three years.[38] By reintroducing low-intensity burning patterns, star thistle can be phased out in the process of enhancing important native species, especially those food, medicine, and fiber crops esteemed by people indigenous to the area.

Reinstating fire management for restoration in the modern day is complicated by private property boundaries and valuable real estate, which are often located at random points throughout an ecosystem. There may be a perfect opportunity to engage in grassland restoration on one property, but if a neighboring house is located on top of the ridge, in the direction that fires would travel, fire management is not a good option. Luckily, strategies based on the ecology of star thistle and where it is found can also be successful in the long term.

Although star thistle forms dense colonies that inhibit germination of other plants, the process of inexorable succession will proceed, ensuring that different species will begin to colonize these zones over time. Though this trend has yet to be researched in areas invaded by star thistle (and is generally under-researched within the literature of invasion ecology),[39] the process of succession will act on colonies of star thistle. No species will grow in perpetuity on a site, especially not in the numbers found in recently invaded sites. Eventually rapidly growing, pioneer species like star thistle will be replaced by longer-lived plants. Star thistle fails to thrive in shade, so as soon as the subsequent canopy layers of trees and shrubs emerge, star thistle populations will die out. A holistic approach to managing star thistle could include planting fast-growing native tree and shrub species, such as buckbrush, mountain mahogany, and redbud – all also nitrogen fixers. When planted within dense star thistle patches, these hardy shrubs will be protected from grazing as they grow the shade cover that will eventually kill the star thistle. Although star thistle suppresses the germination of other annual plants (at least for a time), considering how this characteristic could be used to enhance the growth of longer-lived perennials is a compelling avenue of research.

This long-term approach would enact the permaculture principles of using biological resources, stacking functions, enhancing succession, and obtaining a yield of improved soil and habitat. This type of restoration strategy would also provide for some of the similar ecological functions that star thistle provides, as it is slowly replaced by the native perennial flowering shrubs that build soil and provide nectar, seed, and habitat resources.

Designing restoration projects with the ecological functionality of invasive species in mind helps guide decisions about how to effectively replace them. Simple eradication measures, especially spraying herbicide, don't address the underlying ecological issues that invasive species represent. In Yosemite today, star thistle is treated with the herbicides clopyralid or picloram wherever it appears. This chemical-based approach to managing invasive species misses opportunities to enhance ecological systems in the process of restoring them. It also adds harmful substances to this important habitat and critical watershed.

Holistic approaches, on the other hand, which mimic the ecosystem functions of invasive species, stand a much better chance of success. Careful management and applied ecological understanding can create valuable yields of native plant assemblages and reduced populations of hard-to-manage invasive species without the use of chemicals. Wilderness areas such as Yosemite have a unique opportunity to partner with local indigenous people to renew historic land management practices. Restoration of these cultural relationships would have numerous beneficial social and ecological yields.

Today Yosemite Valley looks very different from John Muir's flower-filled description of the early 1900s. The valley floor is crowded with closely spaced conifer trees, whose dead lower branches create ladder fuels and enhance the potential for catastrophic fires. In 2013, the largest wildfire in the history of the Sierra Nevada Mountains raged through this area, its epic proportions fed by the abundance of available fuel. Another huge fire blazed through in 2014. Though the conifer trees covering the valley floor are technically native, by another standard, they could be considered invasive to this historic agricultural system, since they compete for light, water, and nutrients, are not edible, and provide minimal habitat value compared to other culturally relevant species. Historic management of this area would have inhibited the growth of such trees through low-intensity burns, which favored the renewal and growth of the edible lilies, tarweed, mint, manzanita, ceanothus, spiraea, raspberries, and cherries mentioned by Muir.

Nature and Culture as Cocreations

The scale and extent to which the ecosystems of the Americas were shaped and changed by human hands is difficult to quantify, but modern research increasingly shows that all native plant communities are, to some extent, the products of human intervention. The way people used these species shaped local ecologies and influenced the composition and diversity of the native plant communities we enjoy today. In many cases, what are today known as native plant assemblages are remnants of once intact cultures and the food, clothing, materials for cooking and building, crafts, decorations, and fodder for animals that sustained and enlivened them.

Milkweed was an important native fiber plant used for cordage and rope making throughout precolonial North America. One anthropological record shows that a 40-foot-wide deer net contained around 7,000 feet of milkweed fiber cordage. Milkweed is a perennial plant that used to be managed by fire every one to three years to ensure the healthy regrowth of straight, strong, fiber-filled stems. The fabrication of one milkweed net would have required harvesting 35,000 stalks – about an acre's worth. Ethnoecologist Dennis Martinez states that up to 150 of these nets were used in a single hunt by people living in the central Sierra Nevada region of California.[40]

There were important ecosystem-level effects of the reliance on milkweed as a cultivated fiber resource. It is likely that several thousand acres in California alone were burned and harvested every year to provide enough material for the valuable hunting nets and other uses. One postcolonial northern California botanical assessment recorded that milkweed formed 60–90 percent of the vegetative cover in many areas.[41] The practice of low-intensity burning contributed to the vigor of stands of perennial milkweed plants and also renewed pasture for the grazing deer that the milkweed fiber nets were designed to catch. These animals were used for food, clothing, and tools. The integrated management of important cultural resources (food, fiber, and fodder) yielded an ecosystem uniquely formed and guided to meet people's daily needs.

Using milkweed for fiber also enhanced the diversity of the surrounding ecosystem. One hundred and thirty-two different species of beetles and forty-five different species of bugs visited an experimental patch of milkweed plants over the course of ninety days in a 1978 study.[42] There are at least ten species of insect that feed exclusively on milkweed plants or nectar, including the larvae of monarch butterflies. Many species of insect also use milkweed stems for habitat or hiding, including predatory insects like praying mantids and green lacewings, which also feed on agricultural "pests." In turn, these insects feed numerous species of birds and fish, which feed larger birds and mammals, and so on. Through the management of milkweed as a fiber resource, the ecology of the surrounding land generated biological diversity and abundance.

Careful management of milkweed produced countless ecological and societal benefits. Like other historically tended plant and animal resources, milkweed is more than just a native species. It represents a series of relationships. The milkweed that remains in North America today is a result of actions and decisions taken by people in their efforts to meet their needs over the last several thousand years. In this sense, the cultural need for and use of a particular plant or animal yields an ecosystem that provides for that need.

We need to appreciate the sophistication of historic indigenous land management strategies. Today, most native species are less abundant because we fail to manage them in a meaningful way. In many cases, the decline in stewardship has also provided niches for invasive species to fill. Invasive species are not only a result of modern land use practices but also a result of the absence of tending native plant and animal communities.

In the Great Basin and Northern Plains region of the United States and Canada, cheatgrass is an invasive species of great concern. This annual grass exhibits rapid early season growth, often germinating and growing before native species, such as sagebrush and various perennial bunchgrasses, have a chance to emerge. Because of its rapid early spring growth, cheatgrass sets seed and dries down completely by the time summer sets in, so large stands enhance fire risk in this hot, arid region. Its swift growth also appears to displace native Great Basin species, especially after fire, since the seeds of many plants native to this region require longer germination time, and cheatgrass is up and growing long before other species can do the same. Cheatgrass appears to follow the positive feedback loop of a perfect invader: rapid growth and other advantages increase populations leading to the exclusion of native species.

However, the invasion of cheatgrass and displacement of native Great Basin species is more complicated than the dogged persistence of a uniquely strong grass species. The Great Basin is a vast region of the United States, stretching from the western edge of the Rocky Mountains to the eastern Sierra Nevada and Cascade Mountains and reaching as far south as Mexico and north into Idaho, Wyoming, and Washington. This area was (and still is) inhabited and managed by the Shoshone, Ute, Mono, and Paiute people. The Shoshone tribe was differentiated into different bands, many of which were known by the names of their staple food sources and whose etymology provides insight into the precontact ecology of the area.

For example, bands of eastern Shoshone included the Buffalo Eaters and the Mountain Sheep Eaters, while the Salmon Eaters, Jack Rabbit Eaters, Groundhog Eaters, Wild Wheat Eaters, and Porcupine Grass Eaters comprised the northern bands. The western bands were known as the Bitterroot Eaters, Mentzelia Seed Eaters, Redtop Grass Eaters, Tule Eaters, Pine Nut Eaters, Ricegrass Eaters, Fish Eaters, Ryegrass Seed Eaters, and Buffaloberry Eaters. Though this region today is largely covered by sagebrush, and juniper in the mountainous areas, there are no Sagebrush Eaters or Juniper Eaters present in the historical record of the Shoshone tribe, probably because although these species were present, they are marginally medicinal at best – definitely not staple food crops. The names of these Shoshone bands and the food sources they represent indicate that at one time, the area was much more diverse and abundant in food for both animals and people. And, like most other parts of the non-European agricultural world, these crops were managed with periodic, low-intensity burning that favored their proliferation.

So what happened? The introduction of extensive, unmanaged grazing coincident with European colonization certainly changed the landscape since the herds of sheep and cattle most likely went straight for the wild wheat, porcupine grass, ryegrass, redtop grass, and ricegrass. Then they likely ate the bitterroot, mentzelia, and seedlings of buffaloberry and nut-bearing pinyon pine. They probably even ate the tule. Though sheep and cows did not directly eat the jack rabbits, groundhogs, buffalo, or mountain sheep, they consumed whatever those animals had historically subsisted on (most likely perennial grasses and forbs). The sagebrush and juniper are not favored foods for these grazers, and so they have come to dominate in the region as overgrazing removed the other species. These types of grazing practices have continued essentially unchanged from the time the first herds of domestic animals were brought to the region and allowed to overgraze.[43]

While sage and juniper are technically native, their dominance due to poorly managed grazing regimes yields land very low in biodiversity. In addition, cheatgrass is able to easily colonize the bare, disturbed ground that surrounds overgrazed sagebrush-dominated land. A focus on eliminating the cheatgrass misses the overall picture of how this land has changed over time and how it could be managed for the type of biodiversity that existed when there were people growing, harvesting, and eating the wild wheat, ricegrass, buffaloberry, bitterroot, buffalo, jackrabbit, and mountain sheep. Eradicating cheatgrass from monotypic stands of sagebrush won't restore anything besides an already fragmented ecology.

In fact, if cheatgrass increases fire frequency and decreases the ability of sagebrush to germinate, this could be seen as an opportunity to replant and diversify the fire-dependent native flora and fauna. Experimenting with fire duration, frequency, and seasonality would yield information for controlling cheatgrass and the optimal establishment of native plants. In this way, restoration of Great Basin biodiversity could be achieved by reintegrating the cultural management techniques that first facilitated the historic assemblage of plants and animals.

Although conventional grazing practices facilitated the invasion of cheatgrass, a different approach to grazing management characterized by high-density stocking rates and well-timed herd movements could regenerate brittle landscapes like those found in the Great Basin. Indigenous people maintained these formerly diverse grasslands through the conscientious management of game animals through provision of fresh pasture and hunting. Regular burning facilitated the regrowth of the perennial grasses preferred by buffalo and other grazers. Hunting the animals for food and material needs like skins, bones, and glue kept populations in check. Hunting activities also included managing populations of predators like wolves and cougars, but not exterminating them, since they were valued for their role in herd management as well. Grassland ecologies of the Great Basin thrived with regular patterns of burning, hunting, and herd movement.

Forward-thinking livestock owners practice a similar approach today, known as holistic management or management intensive rotational grazing. Allan Savory, a renowned game manager and biologist, formulated a series of steps that grazing operations can undertake on any scale, which imitates the natural movements of large herbivores like cattle, sheep, buffalo, and wildebeest as they concentrate in large groups to feed, then disperse in the presence of hunters or predators. In a natural setting, these ruminants gather in dense herds wherever they stop to feed. In time, this feeding ground becomes covered in their manure, which is dug into the soil by the action of thousands of hooves concentrated in one place. Although the soil is disturbed and the plants are eaten down, the creatures move on to another feeding place shortly, only returning to the same ground once the grass has regenerated.

These cycles of disturbance and rest promote the even application of soil disturbance and fertility on a landscape scale and ensure that the preferred fodder crops have adequate time to regenerate before the animals come back to feed. Although simulating truly natural herd movement is largely impossible due to private property, any landowner can practice similar techniques on his or her own pasture or rangeland. This style of grazing has numerous benefits, from weed suppression to improved herd health to better soil organic matter.

In contrast, the extensive grazing model largely practiced throughout the world today allows cattle to range freely, eating whatever they happen to come upon as they wander in search of food. In addition, most range managers are notorious in their efforts to eradicate predators, eliminating the bunching and dispersal patterns so critical to maintaining healthy grasslands. The trouble with this approach is that cattle and other ruminants, especially when left to their own devices, tend to eat the most palatable species first, leaving the less-preferred plants. If allowed to wander, they will search out their favorite plants from a broad area. Over time, the overall health and diversity of the range declines since the unpalatable plants have the ecological advantage of not being eaten and are therefore able to reproduce and spread, like the juniper and sagebrush that now dominate Great Basin ecosystems. The quality of the range or pasture diminishes, stocking rates decrease, and over time, the landscape becomes less viable for maintaining either ecological function or economic livelihoods.

The story of rangelands in America has largely been one of declines in soil and plant productivity and animal health. However, by simulating the intensive pulses of foraging and movement found in natural and historic human-stewarded systems, managed grazing encourages soil development and the proliferation of grasslands. When set to graze in this manner, herbivores eat voraciously from the tender tops of all of the plants and then move on before too much living material is consumed. Grazing managers time the feeding so that the plants are able to photosynthesize, rest, and regrow before the herd returns. Instead of decreasing stocking rates, ranchers like Savory advocate increasing stocking rates, sometimes by up to 400 percent. These dense herds are kept in small cells or strips, often managed with easily moved solar electric fencing. Herds need to be moved as often as every twelve to twenty-four hours, which is where the "management intensive" part comes in. But in this case, intensive management also means more opportunities for people to engage in land stewardship practices while retrieving economic value.

Holistic management of grazing animals can also act as a soil-improving carbon pump since every time the growing tops of plants are eaten down, an equivalent amount of roots decompose, slowly but surely increasing soil organic matter. When the timing is right, and the plants get enough time to regenerate, this cycle continually adds valuable nutrients to the ecosystem. In addition, management intensive rotational grazing promotes soils with higher water content, which helps reverse the trend of desertification.[44] Savory and other range managers believe that converting now "unproductive" rangelands into holistic planned grazing systems can even reverse climate change by capturing the carbon cycled through ruminants and storing it as soil humus, which is resistant to further decomposition and atmospheric release. In order to accomplish this in degraded or desertified landscapes, an initial feeding of seedy hay is required, which then sets the stage for plant establishment and growth.

Introducing these types of management regimes in the Great Basin region would improve ecological function, produce better economic outcomes for range managers, increase soil carbon and water-holding capacity, and preserve critical habitat and species. It would create incentives for people to live and work in the area, weaving their lives into the ecology of the region. In a robust and healthy ecosystem tended by people whose lives depend on the health of their surrounding landscape, concern over invasive species could become a thing of the past. Just like in any garden, cheatgrass would be weeded out or outcompeted over time if it is not productive to the system overall. Rather than declines in rangeland productivity leading to eventual desertification, this approach to restoration would provide hope and meaningful work by pairing a healthy ecosystem with people's livelihoods: culture creating nature.

Diversify the Pantry, Restore the Land

One of the best ways to ensure the health of native plant communities is to use them. As we have seen, native ecosystems are more than just aggregations of plants and animals; they are remnant agroecosystems that have evolved through thousands of years of stewardship and engagement. Pantries stocked with corn meal, cattail flour, camas bulb cakes, and water lotus seeds would necessitate landscapes where these species proliferate; insulation made of rushes would make wetlands critical features of home comfort. Saving native ecosystems has everything to do with placing importance on their survival, not as intellectually important vestiges of what used to be, but as vital resources that are integrated into our day-to-day lives.

Imagine if people living in the Pacific Northwest relied on camas, a native perennial lily, for a starch staple instead of wheat or potatoes. Horticulturalist and plant breeder Luther Burbank pointed out that camas is superior to the potato in both "nutriment and flavor" and believed that its culinary use would eventually supplant the potato.[45] Last year, farmers in the states of Idaho, Washington, and Oregon planted and harvested nearly 550,000 acres of potatoes, one of the most heavily fungicide- and pesticide-saturated crops on the market. Modern conventional potato cultivation excludes all other plant and animal species and turns hundreds of thousands of acres of so-called productive farmland into ecological deserts. If camas did replace the potato in our diets, the surrounding landscape would include fields full of long-lived perennial lilies along with numerous ecological associates, including a host of native pollinators and beneficial insects. Camas requires no additional fertilizer or supplemental irrigation and grows profusely without fungicides, herbicides, or insecticides through the rainy winters and heavy clay soils common in the region. The landscape, rather than annually tilled and monotonous potato fields, would be filled with purple-blue blossoms, harkening back to the time when the vast camas fields fooled unsuspecting travelers like Lewis and Clark into believing they gazed on water, rather than land. Meriwether Lewis wrote "The quawmash is now in blume at a Short distance it resembles a lake of fine clear water, So complete is this deseption that on first Sight I could have Sworn it was water."[46]

Undertaking restoration of native plant communities for use by people requires a shift in perspective. Restoration activities that focus on plants at the expense of people fail to make any impact on the very processes that caused declines in native ecosystems in the first place. This is because while well-meaning conservation groups and individuals are busy with the work of "restoring" native ecosystems, the land management systems that contribute to their demise, such as industrial agriculture, mining, and urban development, continue unabated. The paradigms of restoration and the activities of modern civilization are currently separate; they must become integrated. We need to cultivate and eat huckleberries, hickory nuts, camas, pawpaw, and peach palm fruit. Native species must become a part of our lives again.

Many indigenous people believe that it is necessary to engage in meaningful relationships with plants and animals in order to make them thrive. As ethnoecologist Kat Anderson has found, "When elders today are asked why the rich resource base and fertile landscape that they remember as having existed in the past has changed so drastically, they are apt to respond by saying simply, 'No one is gathering anymore.' The idea that human use ensures an abundance of plant and animal life appears to have been an ancient one in the minds of native peoples."[47]

Novel species are bound to invade restoration projects that focus on native species as "wildflowers" rather than food or medicine, because like any garden, they are left unweeded and untended. The plants and animals that are now considered native are not static features of an ecosystem; they are relics of conscientious stewardship, and it is this stewardship that is required if their populations are to be maintained. Many native plants are well adapted to proliferate with the disturbance that harvesting and other cultivation activities provide. As ethnobotanist Nancy Turner writes:

> The culturally important plants . . . are all perennial species, either herbaceous perennials or woody perennials (trees and shrubs). This is the key to the management practices presented here – and to their

> lack of recognition in the past. These plants are not limited to seed production for propagation and maintenance of their populations. They have meristematic tissues in their rootstock, rhizomes, roots, and buds that allow them to grow vegetatively, as well as from seed. The management practices are reliant on the broad capacity of these perennials for regrowth and regeneration.[48]

Camas reproduces both vegetatively and by seed. The edible bulbs are gathered in the late spring when the ground is soft enough to dig. They are usually the size of small onions, and they have a series of tiny bulbils that hang off the bottom of the roots. When the large, fleshy, older bulbs are dug, the disruption dislodges the bulbils and ensures successive harvests. In addition, digging thins the bulbs, providing more space for future generations of plants to grow big and develop their starch content. Other important starch crops, like groundnut, also improve yields and vigor through careful harvesting.

Gathering native bunchgrass seed, such as basin wildrye or wild rice in the traditional manner of beating the ripe seed heads with a wide woven rush paddle ensures that plenty of seed will be left on the ground or in the water for other animals to eat and to germinate the following year. Wapato grows as a tuber in wetland environments throughout the temperate world. Harvesting the wapato involves stomping the mud to loosen the larger tubers, leaving the smaller ones and broken pieces behind to continue to grow.

Bringing It Home

Native plants aren't just native to a place, they are essential elements of a culture that stewards them. Many native species depend on human use for their proliferation. Native species like milkweed, echinacea, trillium, bloodroot, dwarf trout lily, biscuitroot, elderberry, bitterroot, huckleberry, hickory, water lily, and cat's claw – as well as countless others – are not merely native plants or "wildflowers." They are food, medicine, fiber, shelter, tools, and a biological legacy. Vast swaths of the Americas, Australia, and Southeast Asia were managed for perennial pasture and food forest characteristics through guided succession and planned disturbance, which created great biodiversity throughout time and space. With the death and displacement of indigenous cultures in the Americas and elsewhere came a massive shift in ecological composition. Less diversity and abundance of native species is the natural consequence. Today, many of the areas known as "natural" or "wild" bear the marks of both historic human management and the lack of it.

Rather than leaving "nature" alone as mandated by the current legal framework of what constitutes a wilderness, we will need to find ways to insert ourselves into it in meaningful ways. We will have to learn to manage succession and mediate disturbance to create high diversity. We will need to refashion long-standing relationships with landscapes that provide for our needs rather than continuing to rely on far-flung centralized production and distribution. Certain tribal groups in the United States and Canada are managing their sovereign lands in these ways and rekindling their traditional management systems. In 2011, the Yakama tribe collaborated with the US Forest Service in Washington to institute low-intensity burning in what were once prime huckleberry fields managed by their tribe. Indigenous burning in the area was outlawed in 1910, so over a century of conifer growth has shaded out the berry bushes. With the reintroduction of fire comes a reintroduction of a sacred food for the Yakama people, as well as the diverse early successional ecosystem, including edible beargrass, bunchberry, and gooseberries that grow alongside the huckleberries. The proliferation of these edible species provides food for bears, birds, and smaller mammals, and each in turn has rippling effects throughout the food web. Diversity is enhanced many times, making the ecosystem more resilient and less invasible.

The value in cultivating native crops and rekindling historic stewardship patterns is a critical piece of cultural revitalization as well as an acknowledgment, on the part of those who learned and accepted a woefully misrepresented history of colonization, that people were intrinsic to the creation and maintenance of natural resource wealth that was "discovered" and subsequently exploited in North America, Central America, South America, Australia, Africa, and throughout Asia. The land management models of people who have lived in a place and enhanced its biological diversity over the course of thousands of years should serve as inspiration for the new paradigm of restoration.

The processes of renewing relationships of stewardship with native species-based ecosystems do not necessarily entail that people are managing every interaction. There are countless ecological relationships and processes that have intrinsic value just as they are. By using our innate faculties of observation and assessment, we can create conditions for these interactions to take place and, ultimately, for life to thrive. Just as a conductor knows all of the parts in a symphony but does not play every instrument, people as stewards can see how all of the elements in an ecosystem relate. We need to reclaim our place in the functional patterning of nature. We are dreamers and visionaries, and our unique abilities enable us to shape the evolutionary trajectory of life on this planet, something that we have certainly achieved already, for better or worse, in the pursuit of meeting our own needs. It is time to apply this potential not just in terms of creating a livelihood for ourselves, but for all of the organisms with which we share the planet.

CHAPTER 7

Restoring Restoration

"I know of no restorative of heart, body, and soul more effective against hopelessness than the restoration of the Earth."

– Barry Lopez

So where do we go from here? If we're going to solve some of the most pressing concerns that are commonly blamed on invasive species, including extinctions, loss of biodiversity, and biotic homogenization, we must take a deep look at the systems that currently provide for the needs of the majority of the world's population. Invasive species are related to these production systems, but eradicating them does nothing to address the fundamental reasons for their proliferation.

One of the reasons we've been so misguided in our approach to managing invasive species is because managing them effectively requires something far more challenging and more powerful than the business-as-usual approach of aggressive, extensive annihilation of "offending" plants and animals. Effective management requires that before anything else – before we develop a plan or reach for the herbicide – we have to first teach ourselves to think differently. Thinking differently, holistically, about invasive species and their management will take time and creativity, as well as honest attempts to answer difficult questions about the state of the ecosystems where they thrive. But if we are to embrace the true meaning of ecological restoration as repairing degraded ecosystems, then we must move beyond short-term and short-sighted invasive species eradication measures and begin the necessary process of engaging the critical task of restoration on all fronts, from our homes and neighborhoods, to farms, forests, and the economic systems that support them – as well as to restoring our own way of thinking.

The process of incorporating a holistic framework into invasion ecology and restoration is emblematic of the historic shifts in scientific paradigms as philosopher Thomas Kuhn explained them. The old paradigm sees invasive species as bad actors, whereas the new paradigm understands the conditions that support their growth. Instead of engaging in eradication, the new paradigm of restoration seeks to address systemic issues with systemic solutions. The first step in this holistic approach is to acknowledge ourselves as part of a web of relationships in which every action has consequences throughout the ecosystem where we live, from our immediate vicinity to the entire biosphere. Although modern life can make it seem as though our lives are dissociated from natural systems, everything is connected. Engaging in a systems-based analysis of the ecological connections that support our lives requires honesty and courage because it's difficult to admit that some of the things we do are so harmful to life on the planet.

Author and philosopher Joanna Macy has written extensively and eloquently about how acknowledging our collective concern for the dying world can be a means for transforming grief and despair into positive action.

> We have to honor and own this pain for the world, recognizing it as a natural response to an unprecedented moment in history. We are part of a huge civilization, intricate in its technology and powerful in its institutions, that is destroying the very basis of life. When have people had this experience before in our history? We ask people to relate to what they experience with respect and compassion for themselves. They're not just griping and grumping. It is absolutely shattering when we open our eyes and see that we are actually, in an accelerating fashion, destroying the future . . . People fear that if they let despair in, they'll be paralyzed because they are just one person. Paradoxically, by allowing ourselves to feel our pain for the world, we open ourselves up to the web of life, and we realize that we're not alone . . . The response that is appropriate and that this work elicits is to grow a sense of solidarity with others and to elaborate a whole new sense of what our resources are and what our power is.[1]

Macy calls this process "the work that reconnects." It is at once humbling and empowering and informs a more realistic relationship with land. We have to face the facts, heartrending as they may be. We must take stock of the world as we have created it, the garden that we have fashioned, for better or worse, into its current form. Cities and suburbs, slums, sewers, garbage dumps, dead zones, feedlots, shopping malls, and monocultures. Pesticides, plastics, and persistent organic pollutants. Coal and missing mountaintops. Oil and the wars derived from it. Climate change, sea-level rise, cultural and ecological displacement, extinction. All of these are real, and serious, threats to the existence of life on Earth as we know it and into the future.

When placed in this context, the proliferation of a particular plant, fish, or mollusk seems paltry. But engaging in eradication-based restoration provides a clear-cut and tangible physical action that seems to address the dire circumstances we find ourselves in today. Spending a day pulling out salt cedar, teaching children about "Aliens in Our Watershed," or getting a job doing wetland restoration seems to offer tangible results in the midst of so many things that don't. It's reassuring for us to experience those results amid so much uncertainty and powerlessness. Unfortunately, it's a misdirected effort, and the energy, resources, and people power supporting the current restoration paradigm would be better spent engaging in management practices that fundamentally transform the processes that contribute to widespread ecological demise – even if their results are not so immediately apparent.

Management, or lack thereof, is key to understanding what can be done, not only to deal with rampant species, but also to conserve and enhance biodiversity within the context of a changed and changing planet. We need to move beyond the appealing myth of pristine, untouched, virginal wilderness, of passive native ecosystems being overtaken by aggressive invasive species; it's a psychological battle as much as anything else. The old paradigm sees "nature" as static and changing little – which is supposedly when "nature" is at its best and most pure – until humans impose our will on it. This philosophy is wrong on so many levels, but especially in how it negates the healthy relationship that people can develop with the natural world by shaping and being shaped by it, as they meet their needs for food, fiber, shelter, fuel, and medicine. The wilderness concept, a distinctly American idea, denies the work of indigenous Americans in creating the abundance that was "discovered" by European colonists not long ago. The abundant game, lush prairies, bustling rivers, towering forests, and indeed the very structure of native plant and animal community assemblages were shaped by peoples' interactions with them over long periods of time and will require the same if they are to persevere into the future.

Tending a garden is one of the best ways to manifest – to put into practice – this paradigm shift. If you've tended a garden, you are already familiar with the management choices you need to make to produce the desired crops. You thin newly emerging carrots and beets to make room for the strongest and healthiest individuals; you weed garden beds and weed them again to ensure the desired plants have access to the nutrients, light, and water they require to make the most nutritious food. You save seeds from the most hardy, beautiful, productive, and delicious crops. Raising animals involves similar decisions – you monitor pastures closely for growth and species assemblages; you breed and cull herds and flocks over time to create populations best suited to your conditions and desires. These activities place humans in an important role of directing ecosystem succession and evolution.

Although eradicating invasive species (even with herbicide) is technically weeding, the ecosystems that these activities occur within are not treated as gardens. If they were, the purported threats that these organisms pose would be dealt with by people who are observing and interacting with the land around them on a day-to-day basis. Having a presence within and tending a patch of land, no matter the size, means that you are constantly tinkering. If you are familiar with the ecosystem where it is found, even the most frustrating of invasive species can often be dealt with relatively easily, and without herbicides or other short-term eradication measures, with a bit of work and creative management.

But the best reason of all to observe and interact with a piece of land is because it matters to you in a real, visceral way. Your life and the lives of those around you depend on engaging with your home ecosystem. Lamenting the loss of biodiversity and displacement of native species while sitting in a classroom, at a coffee shop, or at the dining room table does little to address them. As the old adage says, "You can't plow a field by turning it over in your mind."

Obtaining a Yield

Although we'll talk about permaculture principles and ethics more in the following chapter, the principle of "obtaining a yield" offers a good example of the kind of psychological shift we need to practice. Rather than thinking of an invasive species as something that deprives an ecosystem and must therefore be eradicated, we must get into the habit of considering what an invasive species has to offer. For example, many invasive species can be managed for multiple yields including compost, animal feed, nectar for bees, medicine, food for people, bioremediation, and the development of meaningful livelihoods. When their management is approached in this way, people steer the ecosystem toward their desired outcome or species profile while creating and harvesting value.

For example, land covered by kudzu could yield several useful products as it is converted into more diverse habitat or productive farmland. Kudzu, a perennial vine in the pea family, or Fabaceae, is considered one of the worst invasive plant species in the world. It covers approximately 7 million acres in the United States alone. It can grow up to 1 foot per day, up to 60 feet per year, and has roots that may reach to 12 feet deep and weigh 200–300 pounds.[2] An impressive species to say the least. As a member of the pea family, kudzu is a nitrogen fixer, and like many leguminous crops, it makes excellent fodder for grazing animals that relish its high protein content. The root, commonly known as arrowroot, is an edible starch that can be used as a natural thickener for soups, candies, and preserves. The flowers are richly scented and favored by bees, which produce a deep purple, grape-flavored kudzu honey. Kudzu provides abundant yields of both human and animal food, fiber, medicine, compost, small business opportunities, and land restoration activities. Managing this plant requires creativity, but there are numerous potential yields, especially since it grows so profusely. Consider this theoretical scenario of kudzu yields:

The plants are rotationally grazed by a goatherd who uses them as high protein fodder for her goats, whose delicious and healthy pasture-raised raw milk she sells at the farmer's market. This provides the goatherd with an income, as well as trading opportunities with a local beekeeper who makes delicious grape-flavored purple kudzu honey, which becomes the sweetener of choice in the local schools, replacing high-fructose corn syrup and refined sugar from genetically engineered sugar beets.

A worker-owned restoration cooperative chops the kudzu vines to make high-quality compost to sell at local garden stores, as well as to enrich their own small market gardens, the produce of which is traded with the beekeeper and the goatherd. An enterprising community member dries and powders the root to sell as part of a dried soup mix in regional natural food stores, as well as to the local acupuncturist, who uses kudzu root as part of her treatment protocol for many ailments, including tinnitus and vertigo. A local restaurant owner experiments with kudzu tofu and steamed kudzu greens on the menu. And a craftperson weaves kudzu vine baskets for sale at local art galleries and garden stores. Another community member makes paper and cloth from the stem pulp, bringing renewable and perennial products into the pulp industry. A local naturopath is inspired by the research study at Harvard Medical School that is using daidzin, a compound found in kudzu leaves, to treat alcoholism.

A couple who buys a farm covered in kudzu decides to try managing it as part of an experimental agroforestry system and works with the beekeeper, the goatherd, the naturopath, the restaurant owner, the artisan, the acupuncturist, the pulp pounder, and the compost maker to time their management so that all of the yields can occur, while their farm benefits from the free nitrogen fertilizer. The overstory crops of high-value native timber and nut trees, such as hickory, chestnut, and black walnut, thrive in the rich soil, eventually growing tall with wide canopies that shade out the consistently mowed or grazed kudzu. The rampant growth of kudzu provides numerous yields as the ecosystem moves toward more desirable ends.

Kudzu and other invasive species can be managed over time in such a way as to eventually deinvigorate their growth and replace their rampancy with other more desirable ecological functions, including the proliferation of native species. In this context, the invasive species is not eliminated just because it is invasive, just as the native species are not emphasized solely because they are native. Instead, these organisms are viewed in terms of the ecological processes, functions, and services that they provide and are managed appropriately over time in ways that make beneficial use of each of these characteristics.

Some invasive species are so productive that they might be able to supplant or even replace the production of other crops or commodities that are today considered "normal" but are actually degrading ecosystems. The tree of heaven provides a good example of an invasive species that could be managed and even cultivated rather than removed. Tree of heaven is considered invasive throughout the United States, New Zealand, and Australia, and its tenacity even inspired a book.

> There's a tree that grows in Brooklyn. Some people call it the Tree of Heaven. No matter where its seed falls, it makes a tree which struggles to reach the sky. It grows in boarded up lots and out of neglected rubbish heaps. It grows up out of cellar gratings. It is the only tree that grows out of cement. It grows lushly . . . survives without sun, water, and seemingly earth. It would be considered beautiful except that there are too many of it.[3]

Tree of heaven resprouts vigorously when cut and spreads well from its prolific seeds. It thrives in high-light conditions and is often found in recently cut hardwood forests and other disturbed areas. It thrives in poor, acidic, and salty soils. It has been used to revegetate mine tailings and tolerates high levels of atmospheric pollution. It is resistant to fumes produced by the coke and coal tar industries and dust produced by cement manufacturing. Tree of heaven accumulates mercury in its tissues, and its leaves concentrate sulfur dioxide, a common component of car exhaust and coal power plant effluent.[4] The roots, leaves, and bark of the tree are used in traditional Chinese medicine for their astringent qualities, and researchers recently found that the plant contains remarkable antimicrobial compounds, which have even been shown to kill strains of drug-resistant malaria.[5]

In addition, *Ailanthus* species provide food and habitat for the ailanthus silkworm, which can survive cold, wind, and drought, making them much more hardy than the mulberry-based silkworm cultivated for the modern silk industry. The ailanthus silkworm (which is technically a larvae that eventually grows into a moth) produces an extremely durable fiber, known as ailantine, which is "more beautiful and much stronger than cotton, and may be regarded indeed as intermediate between silk and wool."[6] The ailanthus moth is incredibly beautiful, with varied colors and a wingspan of over three inches. This fiber-bearing insect/tree polyculture was introduced in the nineteenth century as an alternative to cotton and produced excellent yields of ailantine in Europe, Africa, America, and Australia. But ailantine production fell out of favor as more mechanized techniques of harvesting and processing cotton emerged.

Today The Nature Conservancy recommends "controlling" tree of heaven with 2,4-D. However, given many of the toxic contexts where it grows, it may be worth cultivating it specifically for the multiple purposes it serves – for bioremediation, medicine, and fodder and as a host plant for silk-bearing insects. Making use of and even cultivating an invasive species like tree of heaven should be evaluated in the context of how these needs are currently being met and based on a whole systems approach to the findings.

For example, consider modern fiber production: Around 11 million acres of land in the United States are used to grow cotton. Worldwide, the total is close to 85 million acres. Cotton feeds heavily on soil nutrients and relies on heavy applications of pesticides and fertilizers to maintain adequate yields for the global textile industry. Much of the cotton currently produced in the world today is also genetically modified, which presents a host of political, economic, and ecological problems. (These massive problems are beyond the scope of this book but are covered thoroughly by other sources including in the work of Dr. Vandana Shiva.) In the United States, cotton production drove the country's slave trade, and worldwide cotton farming remains one of the worst types in terms of social and environmental justice. After a spate of well-publicized suicides by cotton farmers in India, several studies found that the combined stresses of poverty, unreliable crop yields, and lack of diversified income streams were among the leading contributing factors.[7] Overall, the transition from subsistence-based agricultural systems toward the production of commodity crops for the global marketplace underlie many of these and other negative social outcomes.[8]

Cotton-growing agricultural systems were instrumental in inspiring J. Russell Smith to write *Tree Crops: A Permanent Agriculture*. Smith describes a scene from the early 1920s, where annual production of corn, tobacco, and cotton had eroded and degraded soils at an alarming rate. Upon viewing landscapes where cotton cultivation had thrived and collapsed, he noted that "the hills are gullied. The fields are barren. Tenantless houses, dilapidated cabins, tumbledown barns, poor roads, poor schools, and churches without a pastor – all are to be found in too many places. No wonder that whole townships are for sale and at a cheap rate. But these neighborhoods might be transformed through tree-crop agriculture."[9] Smith's dire observations took place prior to the Green Revolution's development of industrial agriculture based on synthetic fertilizers and chemical pesticides, which have since artificially enhanced the productivity of such crop production systems. These artificial production inputs will eventually fail, and we may see scenes like those described by Smith as features of the future rather than relics of the past.

In contrast, the cultivation of perennial crops requires far fewer inputs and provides more ecological benefits over time.[10] The Food and Agriculture Organization reports that perennial crops improve and protect soil, increase water infiltration and improve water cycling, have greater potential to capture and store carbon, feature greater above- and below-ground biological diversity, require less labor, and provide for diverse yields and income streams, which enable the potential to redirect labor toward livelihood needs rather than cash production.[11] Taken in this context, cultivation of a perennial fiber-producing plant like tree of heaven instead of cotton is a compelling idea that could provide numerous economic, social, and ecological benefits. Imagine a scene where tree of heaven is cultivated rather than eradicated:

Large swaths of acidified soil surrounding ancient coal-fired power plants are planted in groves of ailanthus. Tailings from mine operations are reclaimed through this tough tree's ability to grow in poor, contaminated, and rocky soils. As the soils are restored and heavy metals are accumulated in the tree's tissues, people in the surrounding area propagate cuttings of other useful tree and shrub species that can follow the ailanthus once its reclamation activities have renewed the soil and made it suitable for other types of plants. An industrious young engineer crafts a chipper that runs on methanol made from composted ailanthus chips. Her designs are fabricated from scrap materials available in the shells of old factory buildings and smokestacks. In a new research lab housed in a former cotton mill, scientists are busy figuring out how to safely remove the mercury from the plant tissues and render it insoluble through a unique combination of anaerobic digestion followed by fungal chelation. They look forward to trading and selling their processes and tools to heavy-metal contaminated sites throughout the world, particularly those using plants and fungi as primary agents of bioremediation. As malaria and other tropical infectious diseases reemerge in the United States as predicted by the most recent models for climate change and vector suitability,[12] extracts from tree of heaven become valuable commodities for people seeking safe, natural alternatives to treating these devastating diseases.

In most communities lucky enough to have flourishing groves of ailanthus, a bustling textile trade is underway. Ailanthus silk is used as the primary fabric for an emerging fashion industry. Lightweight, warm, and durable, this fabric serves the population of hard-working farmers, foresters, surveyors, builders, healers, biologists, engineers, and artists that have sprung up in and around aging industrial centers. In areas full of ailanthus where people are hoping to set up homesteads, herds of llamas and alpacas graze the shoots on a rotational basis, gradually weakening the roots by depleting the plants' photosynthetic surface. Their wool is woven with the ailanthus silk to create a natural alternative to synthetic fleece fabric, and it is rapidly gaining popularity as the warmest, lightest natural fabric in the world. The ailanthus moth, whose cocoons provide the silk for this textile, is the second largest moth in the United States, and soon becomes an icon representative of an emerging sustainable economy based on perennial cropping of textiles in the northeastern United States.

If tree of heaven is better able to provide valuable yields like food, fiber, medicine, or bioremediation than other species (like cotton) cultivated for these purposes, then managing their populations for these yields – and not planning on their eradication – merits consideration. Compared to cotton, cultivating tree of heaven for fiber helps mitigate climate change, reduces pesticide use, remediates heavy-metal laden soils, and has the potential to create economic outcomes for people living in invaded ecosystems. Though it may seem like the relationship between tree of heaven growing along the New Jersey Turnpike and cotton farmers in India is tenuous, understanding the potential fiber yields of the tree brings the destruction of faraway ecosystems and communities in the name of meeting modern fiber needs into focus. This shift in perspective is crucial to understanding invasive species in a new way because it is not just the species in question that is the issue – the issue is considering how that species could be used in ways that make us more responsible for meeting our needs in ecologically and economically rational ways.

Some invasive species may not point to such obvious pathways to creative management, but their presence in an ecosystem can nevertheless offer a roadmap to best stewardship practices. Others may not require management at all and may instead be important anchors of new ecosystems emerging in the wake of climate change. One thing is certain – invasive species are here to stay, and it is up to us to seek as much information as possible as we evaluate their presence and make management decisions. We can continue to rely on the conventional model that seeks their eradication, or we can adopt a long-term ecologically integrated approach.

Holistic and intensive management may be easier to conceptualize on a small lot or individual property, but some of the places where invasive species thrive most intensively are lesser-managed spaces or those that are traditionally unutilized in the context of long-term consistent stewardship in the modern context. These are parks, wildlands, walking trails, ponds, beaches, forests, wetlands, prairies, and deserts. How could such vast and varied tracts of land be treated as though they were gardens? This is perhaps where a new paradigm of restoration can be most meaningful, if not without challenge.

On privately owned land, the owners are free to manage the resources available how they want, for the most part, with little oversight from community or government. A landowner may clear cut the forest, mine the minerals, or develop a perpetually regenerative perennial polyculture depending on their interest. There is no specific incentive to manage land in a more holistic manner, although perhaps public pressure will force changes over time.

In 2012, landowners adjacent to Squibnocket Pond in Chilmark, Massachusetts, sought to eradicate the invasive common reed growing on their property adjacent to the pond with the glyphosate-herbicide Rodeo. The town council voted to prohibit this action, citing a 1990 environmental assessment restricting the use of chemical fertilizers, fungicides, herbicides, and other pesticides within 500 feet of the pond. The landowners appealed and won by arguing that local pesticide regulations were more stringent than the state's – and that the state's law preempted the local ordinance. In response, town residents, concerned about the long-term health and ecology of Squibnocket Pond, voted to exempt the area around the pond from the more lax state pesticide laws. The bill passed the state House of Representatives in November 2014, making it legal for these local residents to enforce what they believed was appropriate for their particular area.

This bill is important in many ways, but primarily because it sets a precedent for public input on the management of private land even if landowners are operating within the scope of the law (which they were in this case, at least, state law). It also set a precedent for a local ordinance superseding state law – direct participatory democracy in action. Landowners adjacent to the pond are still looking for ways to manage the common reed growing in the pond using organic methods, and this represents another opportunity. For example, leaky water from residential septic systems has raised the nitrogen levels of the pond – common reed is known to thrive in high nitrogen environments and accumulate nitrogen in its tissues.[13] So, a regenerative solution for both the pond and the community could include identifying ways that landowners in the area could transition from nutrient-leaking septic systems toward more ecologically sensitive approaches to sewage management such as the Living Machine system designed by biologist John Todd or harvesting the reed to make compost to enrich local gardens. Opportunities abound for designing invasive species management plans that address multiple issues at once in a holistic manner.

Similar approaches can be taken in public spaces where invasive species thrive. Public spaces like parks and natural areas are managed for everyone equally and are often well-loved by people seeking nature, rest, and relaxation within the context of widespread industrial development. Such spaces are also increasingly on the front lines of the invasive species war since there are often no clear imperatives around how such lands should be managed. The desire to keep things looking and functioning in a certain way can drive decisions to use herbicide or other pesticides.

In San Francisco, California, for example, the Natural Areas Program, a division of the city's Department of Parks and Recreation, is responsible for managing 1,110 acres in thirty-two parks or parts of parks. The agency's use of herbicide to manage invasive species is contentious, as is their far-reaching plan to restore native habitat assemblages through eradication of invasive species (including Monterey pine – a pine native to the area a hundred miles south of San Francisco, but considered invasive in the city).

One issue with the plan is that the Natural Areas Program (NAP) currently employs 9 people out of a total 646 employed by Parks and Recreation – meaning that each NAP employee is responsible for managing an average of more than a hundred acres of land scattered throughout the city, including rugged and difficult to access terrain. Caring for this land requires more work than can be accomplished by a handful of employees and is part of the reason that herbicides are considered necessary in the first place. Unless the plan is to use herbicide to manage these sites in perpetuity – which hopefully it's not – what about evaluating opportunities for city residents to participate in the process of designing, maintaining, and judiciously harvesting from the natural areas?

Such public spaces could be managed cooperatively by community stakeholders available through a permitting system similar to that offered by the US Forest Service for wildcrafting and woodcutting. People could make use of some invasive species for food, medicine, fodder, honey, or compost materials. In cases where eradicating invasive species and favoring native or other less prolific organisms is deemed necessary and ecologically appropriate, their management could be taken on by people testing creative ways of encouraging native species over time. Certain invaded areas could be left as experimental ecosystems – living laboratories of ecological evolution in real time, changing in response to the present conditions. At this point, the "restoration" of San Francisco's natural areas to native species is as much a human construction as are areas that are currently filled with blackberries, eucalyptus, nasturtiums, and ivy – or even the teeming metropolis. These areas already require constant intervention to maintain them in the "native-only" state envisioned by the current management plan.

A Restoration Economy

There are so many ways in which our current model of living with and stewarding natural resources is in need of reimagination and restoration, both economically and ecologically. The current paradigm of restoration seeks to remediate some of the symptoms of ecological decline by focusing on returning ecosystems to a former state. By not focusing on the economic models that have caused the destruction in the first place, it misses the opportunity to acknowledge and address the underlying factors. A new paradigm of restoration must address these issues from a whole systems perspective, which means it must also make ecological restoration economically viable – in order to be relevant and achieve long-lasting results.

The southern Willamette Valley region where I live is considered an economically depressed area. Though it was home to a thriving indigenous economy for thousands of years prior to colonization, the arrival of resource extraction–based capitalism upended the ecological, political, and economic foundation of the region. First the trapping and sale of furs from beavers, martens, wolves, bears, cougar, and otters changed the terrestrial food web. Then, gold was discovered in the hills east of town, and the blasting and dredging of sediment changed the stream banks, releasing ore-bound mercury into the water where it cycles through the aquatic food web to this day. The timber economy took hold soon after, and today many people in the area live off of its dregs. There were once nine sawmills in town; now there is one, owned by the transnational timber operation Weyerhauser, that also owns thousands of acres in the watershed.

The timber industry is considered one of the prime economic drivers in the area, yet in all my years walking along the graveled roads that wind through clear-cuts and crisscross the steep slopes around my home, I can count on one hand the number of times I have seen a person working in the forest. I once saw a person operating a feller-buncher machine and have twice seen crews of workers in white vans with barrels of Roundup strapped to the roof on their way to spray the invasive Scotch broom and Himalayan blackberry from recent clear-cuts. This is the sort of personal interaction and relationship that exist in an area of tens of thousands of acres.

Resource extraction built the town, and although there were once abundant jobs, now they are scarce, and 10 percent of the town's population lives below the federal poverty line. However, even though there are few jobs, there is plenty of work, and it is in this transition that economic restoration can partner with ecological restoration. The acreage that currently supports minimal intermittent paid work is calling out for people to engage with it and diversify its ecological and economic outcomes. The "problems" of invasive species in these industrial contexts could be turned into economic solutions.

For example, monoculture Douglas fir plantations could be thinned and valuable shade-grown medicinal plants cultivated beneath trees that grow to one hundred, two hundred, or three hundred years old, with spotted owls in their branches. Or they could be interplanted with coppice-able hardwoods for furniture making and timber framing, which could supply a number of entrepreneurs and local small businesses. They could be burned to favor the oak and camas that once grew there and supply a regional native food economy and robust populations of grouse, deer, bear, bobcat, and cougar. Small diameter trees could be harvested and used to build gabions in dry creek beds, harvesting sediment and holding water so that seasonal creeks would run year round once again, harkening the return of cutthroat trout, steelhead, and lamprey. Food forests and microfarms could dot the hillsides, with people harvesting and making blackberry bramble and Scotch broom compost to feed their fruiting trees and shrubs.

We need radical restoration of people into their home landscapes, reimagining the possibilities of what it means to be native to a place in the modern era. We need meaningful relationships with land and experiments with how best to mitigate, remediate, enhance, and diversify not only the ecosystems that support human life, but also those that support as many other organisms as possible. As we enter into an uncertain climatic future and embrace the possibility that the production systems we have come to rely on may rapidly become untenable, we need to think differently about how we relate to our place. Restoration in this sense is not so much a process of going back, but of moving forward into the unknown, and using our creativity and the tools available to us to create the conditions in which life can thrive. We need to garden and expand our notion of how everything else does also.

In doing so, we will learn how to best encourage the proliferation of highly diverse and abundant ecosystems and manage invasive species in the process. We will not achieve anything of the sort by continuing to eradicate these novel organisms in the vain hope that the ecosystems where they live will be the same as they were at some idealized time in the past. We are here now, on the cusp of the sixth great planetary extinction, with climate change intensifying, and the ways that we relate to the land that sustains us will become ever more central to designing our way through the challenges to come. We need to understand the ecological lessons embedded in the proliferation of invasive species to determine how we can best assist in bringing diversity and abundance to the ecosystems where they are found. From the Berkeley Pit to coal mine tailings, from clear-cut forests to overgrazed rangelands, to the vast monocultures of corn, soy, and wheat as far as the eye can see, we need to see what thrives, and take note, because there are lessons embedded within this ruderal life that point the way forward to ecosystems designed and managed to serve not only our species, but all life on Earth. This is our imperative, and there is no greater or more pressing work to be done.

CHAPTER 8

Putting Permaculture to Work in Restoration

"There is only one soil, one flora, one fauna, and one people, and hence only one conservation problem . . . economic and esthetic land uses can and must be integrated, usually on the same acre."

— ALDO LEOPOLD

Changing the way we think about invasive species might be the most difficult challenge in restoring ecosystems in which they're found. But ecosystem restoration also requires practical design decisions and management practices – not only a change in our thinking, but design and management decisions that *reflect* the change in our thinking. This is where permaculture design – with its basis in applied systems thinking and integration between human and nonhuman needs – has so much to offer.

An Ethical Foundation

Permaculture offers a set of simple ethics that represent guiding principles for decision making that optimize outcomes for people and the ecosystems on which they depend. These decisions and actions should: care for the planet, care for people, and reinvest surplus energy, money, and other resources into regenerative systems. Although they're subjective, these ethics provide benchmarks and a backdrop for any and every design decision. The importance of a fundamental ethical framework from which to guide decisions is critical to the success of managing complex systems, because there are so many different types of decisions to be made and ideas to weigh. In permaculture, a designer's primary ethical commitment is first to Earth and its inhabitants and then to the goals of the project.

Consider, for example, an individual's decisions around managing invasive garlic mustard. If the individual is using ethics as a backdrop for her decision making, she'll encounter a roadblock when she considers using an herbicide. If she has any information about herbicides, she'll know that she'll bear the brunt of its potential toxic effects, as will other people who may come in contact with it. The ethics of caring for Earth and caring for people provide a way to stop and measure your decisions before you act on them.

In terms of the third ethic – to reinvest surplus energy, money, and other resources into regenerative systems – there may be some small flow of surplus toward the individual carrying out the work, even if she's decided to use the herbicide, if she's receiving an income for her work. This income may be reinvested to some degree back into the local economy and community. However, most of the interaction is a one-way stream of money flowing toward the manufacturer, that is probably not spending a great deal of its own surplus within the same community where the herbicides are applied.

In contrast, if this individual instead goes into the forest armed with a harvesting basket and a hand scythe and gathers the edible plants in order to bring them back to a cooperatively owned commercial kitchen where they process them into lactofermented garlic mustard sauerkraut, she will be adding ecological, personal, and economic value to garlic mustard removal – meeting all three ethics. Over time, the seed bank of garlic mustard will dwindle, and by that time, the forest understory will flourish with an array of carefully cultivated high-value native medicinals including goldenseal and American ginseng, and a diversity of native hardwoods will form the eventual canopy layer of a forest managed for late-successional characteristics. Not only do such practices have the potential to enhance native ecosystems, but they also divert resources away from business interests that have little stake in the health and resilience of local communities. This type of restoration model meets the three ethics of permaculture and is the type of model toward which we should strive.

Putting Permaculture Design Principles into Practice

Permaculture ethics provide a framework from which to base decisions, and the principles outline ways to engage the process of decision making – whether on a farm or in a forest, apartment complex, or boardroom – by thinking like an ecosystem. Permaculture design is, in essence, the art and science of learning to see and think in terms of integrated systems. As Bill Mollison notes, "Permaculture is a philosophy of working with, rather than against nature; of protracted and thoughtful observation rather than protracted and thoughtless labor; and of looking at plants and animals in all their functions, rather than treating any area as a single product system."[1]

People think about specific permaculture principles in different ways. In his book *Permaculture: Principles and Pathways Beyond Sustainability*, David Holmgren describes twelve specific design principles, such as "using edges and valuing the marginal." Bill Mollison articulates permaculture design differently in his book *Permaculture: A Designer's Manual*, but he also includes principles such as "working with nature, rather than against it." I incorporate the work of both Bill Mollison and David Holmgren into my own design efforts, as well as the contributions of other systems-based researchers, designers, and land managers, in order to come up with what I feel is the most comprehensive approach to the design, management, implementation, and maintenance of dynamic living systems. For anyone interested in putting permaculture and systems thinking into practice in ecosystem restoration, I recommend further reading of both Holmgren's and Mollison's work, as well as that of C. S. Holling, Howard Odum, Donella Meadows, Allan Savory, Kat Anderson, Fikret Berkes, and the Resilience Network.

Below I describe principles that are especially relevant for understanding invasive species and making plans for holistic ecosystem restoration. I've grouped the principles into four phases similar to the phases you might undertake during the course of a permaculture design, to demonstrate how the principles support the process of making and implementing a long-term management plan. Phase 1 entails taking time to observe and think in terms of the big picture. Phase 2 provides ways to gather information about the site and the species you are working with. Phase 3 includes principles for making an action plan, and Phase 4 deals with implementation, at which point, the cycle begins again, as consistent adjustments are almost always necessary.

Phase 1: Turn on the "Macroscope"

Permaculture starts with the big picture. It is the "macroscopic" perspective, which is integral to the process of thinking in systems. Eradication-based invasive species management tends to focus on the details of the organisms themselves, rather than looking at the ecosystems in which they are part. We should reverse this sequence in order to develop the most holistic understanding of the organism in question.

Objective Observation

In order to manage invasive species in a manner that reflects the ecosystem of which they are part, it is necessary to develop a practice of objective assessment. As we have seen, the rhetoric of invasion ecology tends toward subjectivity in its description of invasive species. It is important to begin from a place that releases ingrained assumptions. This does not necessarily mean that your management plans are going to involve leaving invasives alone, but it does mean that you will develop a deeper understanding of why they are present in the ecosystem, which will lead to a more thorough and long-lasting management plan that is solidly grounded in practical ecological dynamics.

To engage in this process, start with a "beginner's mind" – that childlike place of wonder where everything is new. Even if you are highly trained in botany or zoology and know all the Latin names of every species in your area, imagine for a moment that you know nothing about them and are seeing them for the first time. See them as elements of a system, neither good nor bad, simply present and making a living from the land like every other species in the ecosystem. I know this is difficult because we've been trained to make quick assessments and snap judgments based on visual cues and previous knowledge. But the practice of true observation involves releasing all assumptions. Zen Master Shunryu Suzuki says, "In the beginner's mind there are many possibilities, but in the expert's there are few."[2]

It can be useful to engage other senses besides sight in the observation process. What sounds do you hear? Smells? How does the air feel? Good observation practices should yield more questions that lead to further observation. This process may take time, sometimes longer than conventional frameworks for managing invasive species provide. The permaculture design process recommends taking at least one year of site observation before beginning any implementation. This period of observation allows you to develop a comprehensive interpretation of the landscape you'll be managing. By taking this amount of time for deep observation, the designer is able to see the landscape in all seasons, during storms and peak water flow events, as well as through high heat and drought. This allows for an understanding of how different species present on the site take advantage of fluctuating site conditions.

Permaculture design uses the concept of "sectors" to explain ecosystem dynamics that originate off site but exert an influence upon it. These include sun angles, the direction of storms, flows of pollution from roads or neighboring properties, patterns of how wildlife makes use of the landscape, as well as the directionality of potential disasters like floods, fires, or tsunamis. By taking the time to observe these dynamics and how they impact and interact with a site or ecosystem, you develop a broad understanding of factors at play, which will help you develop sound management recommendations.

Phase 2: Site Assessment

After you've completed a period of observation, the next step is to find as much information as possible about the site, region, watershed, and so on. This information can come from local knowledge, research into historic climate data, online tools such as those available from the Natural Resource Conservation Service and the USDA PLANTS database. This phase also consists of targeted observations that could include marking high- and low-water levels on a wetland or riparian site, taking photographs during different seasons and throughout the restoration process, doing soil tests to determine pH and presence of nutrients or heavy metals, and so on. The following principles and other tools will help with carrying out a comprehensive site assessment.

Know Your History

Permaculture designers always consider the history of a site and the region to inform their understanding of what species and soil and water conditions may be present. This history should include the deep history of how the landscape has changed over geological time. An analysis of site history over the last five hundred years or so is particularly valuable, including recent consequences of grazing, clear cutting, draining or filling, pollutant buildup, and crop history, among other events. This analysis should include the land management practices of the people who inhabited it for thousands of years.

Use Science (and Systems Science)

There is a lot of research on invasive species. Many of the most notorious organisms have been studied in-depth numerous times over. These studies can be useful, especially if their findings are applied to a wider systemic context. Too often, invasion ecology research is framed in terms of simple, causal relationships, which fail to associate findings with other phenomena beyond a specific hypothesis. As we saw in chapter 2, it's not necessarily that scientific studies are wrong, but that their scope is frequently limited and often doesn't provide the whole story behind factors that may be contributing to an invasion. In order to understand the dynamics driving an invasion, it is necessary to peel back the layers of meaning behind research findings.

For example, Japanese knotweed is considered a noxious invasive species because it retains a substantial amount of nitrogen in its leaves and stems. One study found that it retained 75 percent of available nitrogen, compared to native alder's 2 percent and willow's 38 percent.[3] A conventional way of interpreting these findings is that knotweed growing in riparian areas removes nitrogen from the aquatic ecosystem, which alters the basis of the aquatic food web by using nutrients that would normally provide for the growth of zooplankton. This line of reasoning suggests that if native species like alder and willow grew on the stream bank instead of knotweed, more nitrogen would be available for zooplankton and therefore for aquatic invertebrates and fish. Knotweed appears to be robbing these organisms of their basis for survival, so it seems critical to remove it and restore adequate nitrogen to the waterways.

Or could a plant that retains significant nitrogen in its vegetative material be removing excess nitrogen from the water? As we have seen, nitrogen is an essential element, but only in moderation, especially in aquatic environments. Too much leads to eutrophication, anoxia, and death of aquatic organisms. So if plants growing along waterways are retaining nitrogen, it might actually point to more significant problems of nutrient loading in the riparian ecosystem, perhaps due to sewage and septic discharge or runoff from farms, feedlots, golf courses, and lawns.

So removing Japanese knotweed and replacing it with alder and willow might not be the most ecologically beneficial strategy for the stream ecosystem. Instead, it might be more valuable and long lasting to design and plant nutrient-retaining buffer strips along the edges of properties that border the stream in order to prevent excess nitrogen from leaking into the waterway.

Scientific research as a guide to ecological awareness is important. But we have to place studies in the context of systems science if the studies themselves fail to relate their findings to larger processes or phenomena. For example, if research has shown that invasive knotweed takes up more nitrogen from water than native species of willow and alder, one might assume that this would negatively affect the riparian ecosystem because of its impact on the food web. A systems-based analysis, however, might critique the study for failing to include the baseline level of nitrogen necessary for the proliferation of zooplankton and how present levels of nitrogen in the water compare to this reference point. Without looking at these kinds of related questions, it's difficult to assess whether the knotweed is taking out excessive amounts of limited nitrogen resources or removing excess nitrogen that would otherwise be detrimental to aquatic life.

In addition to making use of the existing research, we also need greater collaboration among academic disciplines, more research into holistic restoration practices, and the involvement of more laypeople and stakeholders. How often and how many times every year does giant reed, or Japanese honeysuckle, or Himalayan blackberry need to be cut down in order to slow its growth? How deep should the mulch be that actively suppresses reed canary grass, and how soon until it can be planted with native shrubs? What is the fire return interval that best diminishes the growth of false brome and increases stands of native balsamroot? How long should a herd of five hundred goats graze a 3-acre field of spotted knapweed? All of these questions deserve answers, and publicly funded land grant institutions should undertake this research with at least as much public financial support as they receive from the pesticide manufacturers.

Seek and Apply Traditional Ecological Knowledge

The principles of permaculture design are, at root, a compendium of wisdom gleaned from the techniques and practices of indigenous societies throughout the world, applied to the modern context. There are sophisticated ways of knowing beyond Western empirical science, and these methods of interpreting environmental patterns and processes are valuable sources of knowledge for regenerative land stewardship. In the Americas and other places with surviving indigenous communities, you should attempt to glean knowledge of the historic and current land management practices that contribute(d) to the proliferation of native species assemblages. Throughout history, people have harnessed and managed the successional drive of ecosystems to create functional, diverse, and high-yielding landscapes. The modern ideology of wilderness as pure, untouched nature has resulted in a lack of succession-based ecosystems management and contributes to the idea that invasive species are disrupting natural systems. Understanding the intersections of historic and modern land management practices is an important part of seeing species invasions in new ways.

Consider the ecological ramifications of changes in these land management practices. If prairies were annually burned and grazed by buffalo for thousands of years, how has the transition to monoculture corn and soy affected the native species assemblages? Are invasive species facilitated by this change in land use? In what ways?

True restoration is complicated, nuanced, site-specific, long term, and personally engaged. There are few groups engaging on this level of restoration, but many more could and should. One of these organizations is the Lomakatsi Restoration Project, in southern Oregon. Lomakatsi engages in watershed-level restoration involving selective thinning, burning, and renewing stands of early successional native species in close collaboration with area tribal members. The Society for Ecological Restoration formed the Indigenous People's Restoration Network in 1995, which has been instrumental in involving indigenous people throughout the world in the design and implementation of restoration projects. These organizations are models for what land restoration and conservation organizations could be.

Discover Connectivity among Site Elements and Their Functions

In permaculture design, things like plants, trees, greenhouses, and water tanks are known as *elements* of a site. Each of these has its own unique characteristics and also performs several *functions*, which can be thought of like ecosystem functions. The more functions and connections among site elements, the better. For example, a water tank serves the function of storing water, but it could also be used to store heat as thermal mass in the greenhouse, provide a structure for climbing vines to scramble up, and be used as an element that adds privacy. Its overflow could be directed to a duck pond, which could also grow aquatic plants like water hyacinth that could be used for fodder or compost making. If all of these functions are utilized, the incorporation of such an element into the ecosystem makes sense because it is connected to many other elements and their respective functions. Permaculture design is not just about having a water tank, or a duck pond, or a greenhouse, but rather about how these elements both provide for and are supported by each other.

In highly functional natural systems, every function is supported by multiple elements. In such places, life-provisioning processes or functions, like nutrient cycling or water filtration, are carried out by diverse species, every one doing its part. This way, if one species is affected by disease, fire, or another disturbance, their service to the rest of the ecosystem is taken up by other capable species. Because living systems tend toward complexity, if only one or two species are thriving where there once were many, it's important to look at the underlying ecological factors that may be contributing to the lack of diversity.

In this way, invasive species can be seen as filling in the gaps in ecological systems that are on the verge of severe impairment or phase change – systems teetering on the edge of "becoming unrecognizable" as described by Donella Meadows. They are the essence of *viriditas* – life's tendency to thrive even in unexpected and unforgiving places. Though some species invasions appear to be occurring in relatively "intact" ecosystems, we have to look at these systems in the context of global climate disruption, historic land use, and other factors such as the change in human management. How are these ecosystems responding in the face of these changes? Are invasive species anchors for future ecosystems?

In an ecosystem, just as every life-supporting function is supported by numerous elements that add resiliency to the system overall, every element in a system also has multiple functions. This embeds a species within the environment as a connected and contributing element, rather than an isolated and malignant invader promoting its survival at the expense of everything else. The following questions can be used to develop a sense of how a particular organism relates to its surrounding environment.

Some Questions to Guide Observation of Ecosystem Function for Plants

- What type of root structure does it have (taproot, fibrous root, rhizome)?
- Does it provide nectar or pollen? At what time of year? Are there other sources available at this time in the area?
- Does it provide shade, shelter, nesting sites, or refuge for any bird, animal, or insect?

- Does it fix nitrogen?
- Does it accumulate heavy metals? PCBs? Other toxins?
- Does it concentrate other nutrients or minerals?
- Does it provide a significant source of biomass (organic material)? Think of its entire life cycle.
- Does it provide an edible nut, seed, leaf, root, or fruit for humans or other creatures?
- Does the plant have any mycological associates? If so, which ones? What ecological function do they serve?
- Does the plant serve a role in water filtration and purification?
- Does it help with erosion control or floodwater abatement?
- Does it ameliorate the process of soil salinization or other trends of desertification?
- What is its successional status? Early, mid, late? How can you tell?
- What is its ecological history? Does it thrive in disturbed soil, high- or low-nutrient profiles, sun, or shade?

Some Questions to Guide Observation of Ecosystem Function for Animals

- What part of the food web does it feed in? Is it a secondary, tertiary, or quaternary consumer?
- What organisms feed upon it?
- Is it recycling nutrients? Where do they come from? Where would they end up otherwise?
- Does it accumulate heavy metals or other toxins in its tissues?
- How is it contributing to the succession of the ecosystem?
- What potential yields could it have?

Get Comfortable with Succession and Disturbance

Identifying the successional status of an invasive species, as well as observing how the ecosystem where it thrives is changing and has changed over time, is an important part of seeing an invasive species in its ecological context. A meadow can be expected to become a forest, and a forest can be expected to lose some of its climax species over time to regeneration and freeing up of nutrients, water, and light. The species present at any point in time are part of this process and offer insights into the underlying dynamics of a site. As we saw with the salt cedar growing where willow and cottonwood once thrived on the Colorado River, many organisms are adapted to particular ecological conditions that might include a certain frequency and scale of disturbance. If these disturbances have changed or ceased, then different species assemblages – adapted to contemporary disturbance regimes – should be expected.

Once you are able to identify ecological characteristics of succession and disturbance and how they contribute to observed species assemblages, the next step is to become comfortable with planning for and managing these processes. Studying traditional methods of landscape-level successional management as practiced by indigenous people in Southeast Asia, India, the Americas, Australia, and Africa is helpful because developing a successional understanding requires knowledge of a vast array of perennial plant species, their preferred habitat requirements and lifespan, and their insect and animal associates – information that is typically outside the realm of most agricultural curricula. For example, how often should oak savannas be burned to ensure consistent acorn production? When is a peach palm ready to harvest? How big should the blueberries be when the hickory seedlings are planted? How old should hazel canes be for optimal basket making? These and other questions are important parts of this successional learning process. In many cases, this knowledge is still available and put to use in many indigenous communities throughout the world.

This type of knowledge is critical to the successful implementation of restoration of native plant communities. When I worked at the Lane County Department of Public Works transforming former farmland into a wetland, and I saw that the native species weren't thriving, my first thought was that the disturbance that had occurred there (removing the topsoil, compaction with heavy machinery) was of the sort that most native species were not adapted to. The seeds we sowed were from any number of native plants, and though we put them in their appropriate habitats – some went on wet pond margins, others in the drier upland environment – many did not thrive, possibly because they were a jumble of early and mid-successional species that needed more than just the right habitat to germinate.

In this case, it might have been better to plan on sowing native pioneer or early succession soil-building plants first, letting them do the hard work of decompacting the soil for a few years before other more tender plants were sown or planted on the site. If we had, the site conditions would have been more similar to the natural succession of a wetland ecosystem. In the long term, the ultimate desired plant assemblage (which in this case was mostly flowers and grasses) would also have to have been intentionally managed to keep it in this state.

Phase 3: Make a Plan

Once you have assembled all of the information about your site or the species you are working with, it's time to make a plan. Although restoration planning should be as specific as possible, it is always changeable based on observed site conditions and ecosystem responses to management protocols.

Identify Long-Term Goals

Before embarking on invasive species management, it's important to develop a plan with clear goals. This process will be different depending on whether you are a private landowner wanting to control Japanese knotweed in your backyard or have recently been hired by your town's Department of Parks and Recreation to develop a strategy for invasive species management throughout the municipality. Regardless, it's as important to undertake the planning process as it is to have the plan on hand to follow. The process will help you evaluate the wisdom of your undertaking.

The act of restoration implies that there is a goal for a site that moves it from a degraded state to something more beautiful, functional, and diverse. However, these conditions can be met by any number of species assemblages, so it's important to define both the desired end state as well as the management plan that ensures participants will skillfully and consistently guide the project toward this goal. In order to make a comprehensive plan, consider the following questions:

- What is the goal (short, medium, and long term) of the site? Think in terms of five, fifty, and five hundred years. Is the site destined to be a park, a farm, a productive forest, a combination of all of these, or something else?
- What is the management plan? Fire, mowing, hand weeding, grazing, mulching, planting? At what intervals and in which seasons?
- Are there local codes or considerations to take into account? Examples include burn bans, setback requirements, land use laws, wilderness

regulations? How does your plan work with or influence changes within this sphere?
- Who are the stakeholders? Have you consulted them and incorporated their thoughts, concerns, and opinions?
- How can the characteristics of the invasive species be managed to move the site toward the stated goals? Can they be used for compost or mulch material, forage, pollination, habitat, bioremediation purposes, or economic opportunities?
- Is it possible to leave a patch of the invasive species for long-term observation and learning?

Take the Long View

Although invasive species are sometimes assumed to be detrimental because they can grow rapidly, apparently free from the natural "checks and balances" of predation or disease that acted on them in their place of origin, this assumption doesn't usually hold true over time – there are always new relationships emerging as organisms figure out ways to adapt to new ecological conditions, including invasions. Evolutionary forces are always acting on both the invader and the invaded organisms. Taking the long view on species invasions means looking deeply at emerging symbiotic relationships before taking action. Eradication or other control strategies should be measured against the likelihood of invasive species moving into a state of dynamic equilibrium in the ecosystems to which they are introduced. Though we often blame invasive species because they can co-occur with extinctions, their associations are largely correlative rather than causative. Invasion is a natural and predictable way that ecosystems change over time. Given the ongoing, imminent, and unknown effects of a changing climate, these factors must be taken into consideration before engaging in an eradication protocol.

Maximize Diversity

As you consider the long-term plan for your site, make plans for maximizing diversity. Diverse ecosystems are more likely to be stable on a community level as diverse organisms can act in different ways to compensate for fluctuations in ecological conditions, thereby ensuring the continuance of ecological function.[4] This doesn't just mean adding the greatest number of species (though that is part of the consideration), but focusing on understanding and maximizing their functional relationships so that the system as a whole is more durable in the case of perturbation.

Permaculture designers act on this principle in a few ways. First is through the concept of edge, which is an ecological term for the interface between two different habitat types, like the edge between a field and woodland, or between a lake and a forest. Ecological edges tend to be more diverse because they contain organisms that live in both the field and woodland, as well as some that thrive particularly in the unique microclimate of the edge zone itself. Many invasive species gain a foothold in unmanaged ecological edges, like roadsides and field margins, because these spaces provide unique conditions for them to thrive and they are outside the scope of management goals that achieve the desired outcomes – such as maintaining car access or retrieving adequate crop yields. From these places, invasive species can move into even seemingly intact ecosystems. Researchers have observed the "edge effects" of clear cutting or tilling – such as moving dust and hot dry winds – up to 500 meters inside an undisturbed forest.[5]

However, ecological edge can also be considered a diversity-enhancing feature in a landscape. Many permaculture plans call for increasing edge within a site by providing vertical structure in the form of trees, vines, and shrubs that help moderate wind, trap heat, and provide habitat for birds, insects, and small mammals – making a landscape much more biologically diverse than if it had been, say, a lawn. This diversity in structure and function also fills niches that may otherwise be filled with invasive species working toward the same ends.

On a large scale, increasing edge by encouraging many different types of successional stages, such as the shifting perennial polycultures maintained by the Yucatec Maya and other indigenous groups, contributes to increased regional biodiversity by creating a landscape mosaic in which many different types of species, from the closed-canopy forest dwellers to those that relish open fields, can proliferate.

However, too much edge can be detrimental. In parts of the Amazon basin where forest clearing is rampant, some species that depend on the presence of unfragmented habitats for their reproduction are in peril. In these areas, forests now exist in a series of fragmented islands, and unfortunately rather than being managed as a shifting mosaic of constantly growing perennials, herbs, and other multifunctional species, these clear-cut forests are transformed into rangeland for cattle, fields of soybeans, and, increasingly, oil palms. These edge effects should be considered when making a plan for restoration. In small islands of natural areas surrounded by cities, they are likely to significantly affect the plant and animal composition in perpetuity. Edges can be used to an advantage though, as they provide unique habitat. On smaller scales, these are the places to focus on first, creating buffer zones where all the niches are full and no resources are available for invasive species.

As with many things, when it comes to maximizing diversity, there is a balance between lack of management and too much. This principle is also related to developing a keen understanding of how to guide succession, as well as thinking about diversity on a regional or watershed scale and over time. It's important to emphasize that maximizing diversity doesn't just mean adding as many plants or animals as possible, but seeking the greatest diversity of functional interactions among elements so that opportunities for invasion are planned rather than unexpected.

Use Appropriate Technology

A key part of developing a holistic plan for managing invasive species is identifying the appropriate tools and practices to use in the process. Permaculture-based planning is based on the principle of using appropriate technology, which as a subjective term may mean the application of different approaches depending on the scale and setting of the project, but the overall practice should always meet the three foundational ethics. For example, the management plan for high-biomass invasive species like giant reed, kudzu, Japanese knotweed, Scotch broom, autumn olive, and even spartina might include cutting with loppers or machetes and using the material for building compost.

One of the ways that I have managed a patch of Japanese knotweed is through cutting its canes every year (after blooming, so I also get the honey) and using them for the base layer of a compost pile. Their hollow, air-filled stems make a great underlayment for a large compost pile in which air circulation at the base is critical. The knotweed stems also decompose over time, which perforated PVC (used for the same purpose) does not – at least not in ways that are ideal for food production.

Such biomass could also be used to build gabions if the plant is located in riparian areas, as these provide valuable sediment-retaining structures in degraded stream channels. Or it can also be used for mulch as part of a more comprehensive planting plan. Sheet mulching is a technique commonly practiced to establish permaculture gardens and forest gardens and entails piling large amounts of biomass (usually at least 18 inches thick), like leaves, straw, or wood chips on the ground to smother the grass or other plants. Eventually this material breaks down into organic matter–rich soil, and you can plant perennials, which benefit from the increased moisture, which also creates habitat for beneficial bacteria and fungi and functions to suppress growth of grass or other weeds. This approach can also be thought of in terms of enhancing succession since it's similar to what a mature forest does over a long period of time.

Any one of the rampantly growing invasive species I mentioned in this book could serve this purpose, although cutting the new growth of such rampant species will invigorate its growth – for a time. Regular, consistent removal of photosynthetic surface will eventually result in the roots being starved of the sugars they need to grow. My observation is that cutting blackberry vines at the peak of flowering, when they have the most energy resources dedicated aboveground, for three years reduces the ability of an individual plant to grow.

Hand pulling and digging can also work, but you need to consider how this disturbs the soil and how that might make it more vulnerable to invasion. For example, digging up Scotch broom disturbs the soil, and Scotch broom thrives in disturbed soil. Its seeds can lay dormant for up to ninety years and are more likely to sprout if the soil is overturned or moved around. A better choice for this and other species that thrive in disturbed sites is cutting or mowing and increasing shade through managed successional perennial plantings.

Many restoration projects use cutting or mowing as part of their management repertoire. When it comes to mechanized technologies, weed whackers typically cause the least impact (a good thing in this case) since they can be used on uneven terrains and are handheld, so they don't disturb the soil. Using brush hogs or other mowing equipment is not as useful in remote areas because of their size. They also compact the soil, so you'll need to consider and plan for this as well if you plan to use this type of equipment as part of a management protocol.

Fire is another tool that can be used to manage invasive species and restore and reinvigorate native plant communities. But the use of fire requires planning and skill. People who use fire as a management tool need to take wind speed and direction, rainfall patterns, firebreaks, and fuel loads – as well as property boundaries – into consideration. Some restoration organizations, like the Lomakatsi Restoration Project in Oregon, use fire, and have expertise in its appropriate regional application. Similar organizations should follow suit. When used to manage invasive species, land managers must take care to determine how the target organism and its seeds/propagules respond to fire. Some respond by growing more vigorously, so future burns or other management methods will be necessary to keep their populations in check. Fire directed specifically at invasive species would probably work best in annuals or short-lived perennials with little to no woody growth.

Backpack flame weeders like those I used on the Lane County wetland mitigation project are also a viable nonchemical approach. In terms of appropriate technology, propane is not a renewable resource, so this option, while more appropriate than an application of herbicide, should be designated as a "quick fix" solution. It does not address some of the underlying issues surrounding the design and implementation of the restoration project itself, but it is an option that restoration professionals could begin using relatively quickly. Timing and fuel load considerations are also important since burning should be done only when conditions are right (low wind, high relative humidity, and early in the day, for example) and all proper safety precautions are in place. In some cases, establishing a simple "fire break" by irrigating around the proposed spot-burn location will suffice.

Another option on certain landscapes would be the use of the Keyline plow, which aerates soil with minimal disturbance. Keyline plows are often associated with pasture and rangeland improvement, where they are used to break up compaction caused by long-term improper grazing. Their slender, bladelike shanks penetrate up to 2 feet deep, creating channels that air, water, and roots can travel through. If drawn across the landscape along contour, water, sediment, and nutrients flowing downhill will run into these channels, decreasing runoff and increasing infiltration. Results are almost immediate since existing vegetation gains access to soil profiles from which it was previously restricted, leading to rapid growth.

Keyline plows have not been used extensively for invasive species management or restoration specifically, although I imagine they would be helpful in certain settings, especially in the establishment or restoration of prairies, savannas, and meadows. Their ability to decompact soils would quickly do the job that thistles and knapweeds perform. Sowing native seeds into the furrows would further assist the emerging ecosystem. Tractors usually pull Keyline plows, though horses or oxen could probably manage a single-shanked plow. The argument could be made that this petroleum-driven machine is not appropriate technology; you'd need to weigh this against the degree to which you think it would achieve vital earth repair work relative to time and resources. If properly managed after establishment, a few passes of a Keyline plow could steer the site onto a path of regeneration and abundance in perpetuity.

Animals are another technology to consider within the framework of restoration. Animals like cows, sheep, goats, chickens, geese, ducks, and even pigs could have a functional role in the implementation of restoration projects. Each of these animals has characteristics that when managed appropriately, can decrease energy inputs while increasing the fertility and diversity of the site. Pigs can be used to root out invasive species like Himalayan blackberries and pampas grass, and goats can browse thorny thistles and shrubs like glossy buckthorn and Scotch broom. Cows and sheep can graze invasive grasses, and chickens can scratch newly germinating seeds. Animal-based management requires a keen sense of timing, good fencing, and/or skilled herders, as well as a plan for providing for their other needs (shelter, water, and supplemental feed, for example). The practice of holistic management has many tools to offer when it comes to restoring ecosystems through the use of animals.

Use Biological Resources

When you're considering how to manage a particular invasive species or engage in ecological restoration of a site, it's worth thinking about how you can use living materials or processes to accomplish your goals – as living systems are the most appropriate technology of all. For example, consider grazing animals before you decide to use a mower. The same task of cutting back biomass will be accomplished, but using animals as a biological resource conserves energy and can also contribute to a yield of meat, manure, milk, fur, or fleece. If mowing is the best option, the cut biomass can be put to good use as compost or mulch, recycling it back into the system from which it originated. Think about the landscape as a living system and consider how your management goals can both utilize and create biological assets.

Although I'm not aware of many restoration projects using this kind of strategy, Lani Malmberg's Ewe4ic (pronounced *euphoric*) Ecological Services based in Colorado is a great example of the kind of ingenuity this can involve. Malmberg, her two sons, two border collies, and 1,500 goats work throughout the intermountain west and prairie states doing weed management, fuel load reduction, and land revegetation for federal, state, county, city, and private landholdings. Malmberg received a master's degree in weed science from Colorado State University, and at the time, she was the only student whose research was not funded by a pesticide manufacturing company. She researched nonchemical approaches to managing spotted knapweed and applied her background in animal husbandry to create her unique, successful, and ecologically responsible restoration business.

Today, Malmberg and her herd of browsers can eliminate a field of Canada thistle within a matter of hours. The goats relish the plants and focus on eating the tender buds first, depriving the plants of their ability to propagate through seed. Thousands of sharp hooves trample the vegetation, and the manure they leave behind enriches the soil. The combination of mild disturbance and manure-based fertility stimulates germination of native grass, forb, trees, and shrub seeds lying dormant in the seed bank. The goats may have to return to the site a few times to clear remaining thistle, but since they are rotationally grazing sites throughout the intermountain west, this is a matter of scheduling the appropriate time based on the optimal regrowth of the plants. Malmberg and her goats leave thriving, diverse, and healthy forests and grasslands in their wake. Rather than focusing on the immediate removal of invasive species, Malmberg's approach is to enhance an ecosystem, developing its resiliency and decreasing the likelihood of reinvasion. Malmberg's model is an inspiration for a new paradigm of restoration. She is developing strategies and protocols outside of industry standards and is working toward long-term restoration at the sites where she works.

One concern with using animals in restoration is the potential for ecological damage through unmanaged, extensive grazing as described in chapter 6. Using animals to restore landscapes requires firm and sensible management, and their feeding and trampling habits require consistent monitoring. Animal manure also enriches the soil, which can bring about ecological changes through the addition of nutrients. Although this influx of nutrients is often considered a "bad" thing in the context of restoration because it influences the successional trajectory of an ecosystem, as we have seen, the state to which an ecosystem is tending is never known. This becomes a matter of actively guiding succession – an opportunity for restoration professionals to delve into and learn how to direct it. We shouldn't be afraid of ecosystems tending toward complexity, instead we should feel comfortable diving right in and lending a hand to the process.

Consider Relative Location

One of the principles used to explain how to maximize the functionality of integrated site elements and relationships in a permaculture plan is the concept of relative location. This principle means that in order to conserve energy and maximize efficiency, desirable elements should be placed in relative location to each other and to their use. For example, it makes sense to place something that you visit multiple times a day, like a bed of salad greens, as close to your kitchen door as possible so you don't have to walk very far to tend and harvest from it.

Permaculture-based planning uses the concept of zones to describe the sensible placement of site resources relative to their intensity of use. In a home landscape, Zone 1 is the area closest to the home, with intensive gardens, greywater systems, and so on. Zone 2 is where your chickens might go, as well as some small fruit trees and berry bushes – things that you visit regularly, but don't need right next to your house. Zone 3 encompasses areas for larger trees, larger animals, and staple crops. Zone 4 could be a place for firewood production and nut trees, somewhere that you really don't visit more than a few times a year. Zone 5 is the less-managed zone, used mostly for contemplation, reflection, and observation.

A comprehensive permaculture plan will feature all zones in close proximity to every home, apartment complex, subdivision, and town to model the intensive, rather than extensive nature of this design system. For many people right now, the potential products of Zone 1, like salad greens, are grown thousands of miles away from their front door. This represents an extensive (though resource-intensive) system. These zones can be designed for the smallest apartment balcony garden to subdivisions to large-acreage properties, and they can even be used on the scale of regional planning.

Imagine the possibilities for comprehensive Zone 5 habitat corridor planning and implementation if the majority of things people need were produced near our homes – leaving room for nonhuman life to thrive and, if necessary, to migrate as habitat conditions changed. We could start by planting gardens in every yard, orchards along every sidewalk, and keeping goats, chickens, or ducks in every vacant lot. Yields of such small-scale, mixed-crop intensive garden systems have been shown to be 20–60 percent higher per unit area than the same area used to produce a single crop.[6] Adopting such strategies would create more space for native plants and animals while reducing the energy and resource consumption entailed in producing our needs far away.

These ideas can be applied to existing urban, suburban, and rural development as well – it's already happening in urban centers like Detroit, Michigan, and Milwaukee, Wisconsin. In Milwaukee, Will Allen's 3-acre Growing Power farm employs sixty-five people in the production and distribution of one hundred thousand tilapia, bees, ducks, turkeys, goats, eggs from five hundred laying hens, and thousands of pounds of fresh organic produce while turning 20 million pounds of landfill-bound food waste into compost and heat for the farm's extensive heated greenhouses. Detroit's Future City plan includes farming in the urban core; biological storm water retention and treatment features; forest, meadow, and riparian corridors; and transformation of roads into rivers. These examples of remediation through intensification are those that will be the future of smart development. The implementation of these and other similar models of human settlement will facilitate the creation of suitable contiguous habitat corridors among their many other beneficial results.

This principle can also be applied when considering how the historic range and dispersal patterns of plants and animals are shifting in response to climate change and habitat modification, as the "relative location" of where we expect organisms to live is changing. Many novel species are thriving in the places where they now find themselves, some, like the Asian carp, more successfully than in their native range, suggesting that we should consider how assisted migration might help mitigate Earth's sixth great extinction.

We need to respect the fact that certain plants and animals are important to indigenous groups and efforts should be made to maintain critical populations of these plants and animals in their native range. However, if ecological upheaval is as significant as predicted by current climate change models, it is going to be difficult, if not impossible, to predict which organisms or species assemblages will survive in the ecosystems of the future.

I like to find plants from within three climate zones in each direction and plant them in my yard and propagate the ones that survive. I live in the Pacific Northwest, so for me this means taking cuttings of cold and blizzard-adapted elderberries from Yellowstone National Park in Wyoming, along with drought-tolerant goji berries from northern New Mexico, as well as many others. I also grow plants from analogous or similar climates from around the world. Coastal Chile, Japan, New Zealand, and the Mediterranean region are similar to the Pacific Northwest, so experimenting with plants from these regions might help to preserve them. Not all will survive the soils or climate where I live, but some will, and over time, my property will become a hotspot of biodiversity from which I can share seeds and propagation materials with others, either regionally or further afield, in order to test what survives.

Of course, a single person or family cannot conserve all the world's vegetative biodiversity on a single property, but a concerted effort among many people from diverse climates will help. This type of active genetic conservation will also help ensure that these plants are better adapted to changing climatic conditions than those whose seeds or propagative material is stored in vaults or herbariums. Actively growing plants are constantly responding to prevailing conditions, and their offspring, especially after successive generations, are better suited to an uncertain and varying climate than seeds kept in storage. (Seed storage is still an important practice for conservation of global biodiversity, but it is better to maintain substantial actively growing populations, especially of rare plants.)

Be Realistic: Quick Fix-Retrofit-Ultimate Permaculture

As you make your plan, it is important to be realistic about the outcomes you should expect. Permaculture consultant Tom Ward uses the concept of "Quick Fix-Retrofit-Ultimate Permaculture" to contextualize different stages or strategies for rehabilitation efforts. The quick fix option is the easiest and most tangible. It's the kind of approach that will do for the short term but may not be as useful or functional in the long term. This is like installing thermal curtains on single-paned windows to discourage heat from leaving the house on cold winter days. The retrofit option gets closer to the root of the issue, but is often more complicated, expensive, resource intensive, or time consuming. Instead of relying on the quick fix of thermal curtains, you retrofit your windows with double or triple-paned glass. Ultimate permaculture is redesigning the house so it takes advantage of passive solar energy stored in high mass walls to warm a greenhouse that recycles all your greywater while growing your winter vegetables.

This is a helpful way of thinking for idealistic designers (like myself) who tend to envision the goal of ultimate permaculture, in which human life support systems seamlessly integrate with and enhance surrounding ecosystems, but which may never realistically come to fruition. It's especially useful in permaculture-inspired restoration and invasive species management, because it offers different scales of restorative activities. For example, a restoration plan could entail pulling out ivy and planting native forest understory species like Oregon grape and sword fern. This would be a relatively quick fix option for the site – the invasive species is removed and desired species are planted. However, this process does not address the larger picture of why the ivy is there in the first place, so it fails to ensure that the ivy will not come back. A retrofit option for the restoration of an ivy-infested forest could be to reintroduce seasonal low-intensity burning, which would require coordination with adjacent landowners and local authorities and likely necessitate years of trials to determine the best sequence of management. The ultimate permaculture option for this site would be for all adjacent landowners to participate in collectively managing the space for their food, medicine, and firewood needs, through the use of fire and other methods of guiding succession that promote high diversity and achieve valuable yields.

The Problem Is the Solution

There are many creative ways to approach the management of invasive species, and they are best when synchronized with deep restoration, which entails the creative redesign of the industrial production and chemically based agricultural models that have so dramatically modified the planet's ecosystems. We have explored many ways that invasive species can be used to move ecosystems toward more desirable outcomes, including using nitrogen-fixing kudzu and Scotch broom as mulch and compost to establish long-term perennial plantings and developing markets for edible species like Asian carp, autumn olive, and sea buckthorn.

Consider how the traits of the species you plan to manage could be used to solve the apparent problems you observe in the ecosystem or elsewhere. As we have seen, the rampant growth of such organisms is indicative of systems and their feedback processes. That they are proliferating can be seen as a positive thing for the ecosystem if we can shift our perspective to see them in such a way.

For example, the rampancy and hardiness of invasive species could be viewed as a potential boon to the development of new varieties of food crops. At The Land Institute in Salina, Kansas, Wes Jackson and other researchers have been working to develop high-yielding perennial grain and legume crops that can be grown in prairielike polycultures. One of their promising developments has been crossing annual grain sorghum with the invasive Johnson grass, a hardy rhizomatous perennial. Drought-tolerant sorghum is grown on 18 million acres in the United States every year, mostly where it is too hot and dry to grow corn, and is mostly used to feed cattle. As climate change intensifies and the plains states become increasingly arid, a perennial sorghum could become an important crop. Johnson grass is reviled by many farmers who see it as a weed, but its capacity to withstand drought, penetrate deep into the soil for water and nutrients, and not require tillage make it a good candidate for creative breeding endeavors.

Another important feature of this and other perennial crops under development at The Land Institute is that pigs, chickens, and ruminants can graze them – something that has the potential to completely change the state of animal production in this country. Currently in the United States, we devote vast tracts of land to annual crops like corn and soy in order to feed the animals we use for meat, milk, and eggs. These crops are harvested, stored, and transported, at great ecological and economic cost, to confined animal feedlot operations. If those animals were instead grazing in high-yielding perennial prairie-based ecosystems, we would solve many of the problems associated with both annual agriculture and animals who live and die in confinement.

Anticipate Slow Variables

This principle comes from the Resilience Network, which offers a series of insights about how to manage complex socioecological systems. Anticipating slow variables means building change, or the ability to change, into your plan. The plan you make for restoring and managing an ecosystem over the long term must have the capacity to adapt to current conditions. Resilience comes about through acknowledging, anticipating, and incorporating constant change in your plan.

Phase 4: Implementation

When it's time to start implementing your plan, there are a handful of other principles to consider, from the philosophical to the practical. These will help develop a sophisticated, long-term approach to regenerative restoration.

Wise Resource Use

Reimagining restoration entails a shift in perspective around people's ability to judiciously interact with and use natural resources. Wilderness ideology restricts people's access to relating with land because it is based on the concept that people always degrade the landscapes in which they live. This may indeed be the case in a society mired in the murk of modern industrial capitalism, such as ours is, and we can be glad that some wild places have been preserved from activities that extract resources, concentrate wealth, and operate as though economics are separate from ecology.

However, as we have seen, there are numerous cultural traditions that demonstrate how direct interaction with and judicious use of resources from the surrounding ecosystem enhances its diversity and resilience on small and large scales. Becoming comfortable with this type of adaptive management and keeping it separate from the dominant economic model will potentially be difficult to overlay in the modern context. Here again is where the permaculture ethics come in handy, since resource extraction purely for private gain does not meet any of the ethical foundations and thus must be reimagined in practice.

In order to begin the process of learning to implement such practices, consider the "Resource Use Rules" compiled by Bill Mollison, in his book *Permaculture: A Designer's Manual.* He ranks resource use practices from most regenerative to most degenerative and notes that resources can:

- increase with use
- be lost when not used
- be unaffected by use
- be lost by use
- pollute or degrade systems with use

Traditional management practices tend to exhibit ecologically regenerative traits, whereas many modern land stewardship practices tend to degrade ecosystems. Too much use (read: nonregenerative extraction) and pollution tend to follow in the wake of current industrial production models. Restoration, in the true sense of the word, should engage only in regenerative practices.

Obtain a Yield

As we saw in chapter 7, obtaining a yield means that the designer intends to produce something from the site, both for herself and for the surrounding ecosystem. We have to begin to see and experience ourselves as producers rather than consumers. Consumers are passive recipients who buy products that show up at a store. Producers actively line the shelves or trade specially produced items among each other. Our current economy is based on consumption and relies on us acting like consumers disconnected from our ability to produce things that will meet our own needs. Bill Mollison relates that "the greatest change we need to make is from consumption to production, even if on a small scale, in our own gardens. If only 10% of us do this, there is enough for everyone."[7]

Transitioning from consumption to production would be a sea change for our economic system. The Greek root of our word "economics" means the management of household affairs. Restoration and other earth repair activities must be tied with economic outcomes, with keeping our "house" in order, with the production of multiple yields that at once serve the needs of the steward and of the ecosystem as a whole.

As we have seen, most invasive species provide numerous potential yields, from habitat to biomass and nitrogen, to food, fodder, and nectar. On my farm, we have abundant patches of Himalayan blackberry. I tend a few beehives, and the flowering blackberries provide a large portion of the bees' midseason nectar. The thorny brambles grow in profuse tangles up to 10 feet high that should only be ventured into with thick jeans and sturdy shoes. I recently found a nest buried deep within a midden of dead blackberry canes, and I have also seen quail and rabbits use the thickets that the canes form for shelter and protection. Many animals, including raccoons, bears, chickens, many types of birds, yellow jackets, and humans, love the fruits. Their sweetness is the quintessential taste of late summer in Oregon.

When we moved here, instead of mowing all of the blackberries, we planned a blackberry wine berry picking party – also giving us an opportunity to meet our neighbors – the first year (to be followed the next year by a blackberry wine tasting party, of course!). We tasted each patch to determine which one had the earliest, sweetest, and juiciest fruit. As we developed a management strategy for the property as a whole, we introduced pigs to root and trample the unwanted patches, turning the patches into islands of fertility awaiting the plants that we would later choose. Pigs love blackberry roots and grow fat as they turn them into winter food for us to enjoy as well. Since we were taking away food, nectar, and habitat sources for the birds, bees, and other creatures, we knew that our planting plan needed to include how we would continue to meet those needs with other plants that would be in line with our long-term management and yield goals of the site. Where the pigs removed the blackberries, we decided to plant a hedgerow of linden, crabapples, hawthorn, spiraea, and red stem dogwood for bee fodder and habitat for birds and other critters. Blueberries, gooseberries, tayberries, raspberries, chestnuts, and wine grapes will eventually grow on the slope the blackberries once covered.

In this way, we are observing yields and making plans based on a plant's usefulness to ourselves as managers and to the long-term integrity of the ecosystem. We obtained yields of blackberries, friends, blackberry wine, honey, compost, pollinators, bird habitat, food for wildlife and domestic animals, pork, soil fertility, and site preparation for future plantings all from one maligned invasive plant. Thinking of invasive species in terms of their yields places them within the ecosystem and encourages creativity around their management. And, as Bill Mollison noted, the yields of such well-designed and carefully managed ecological systems are theoretically unlimited.

Use and Value the Marginal

Learning to value invasive species and ruderal ecosystems as repositories of information and demonstrations of ecological processes is one of the most meaningful actions citizens and professional restoration ecologists can take. We live in a changed and changing world, and those organisms that are thriving in the most challenging or altered conditions provide unique learning opportunities. From the proliferation of *Euglena mutabilis* in a lake so toxic that no one thought life would thrive in it ever again, to the tree of heaven sprouting up in abandoned parking lots, life thrives in unexpected places. Understanding that each organism represents a functional piece of an ecological system, and therefore has something to offer, is critical for beginning the process of holistic restoration. We need to embrace rampancy and consider the phenomenon an opportunity to deepen our ecological awareness.

Using and valuing the marginal also invites cross-disciplinary strategies that can provide creative solutions to challenges we face in ecological restoration. For example, one restoration project I am familiar with is a "nature park" that features a walking path and fishing ponds in a unique floodplain ecosystem. It is also habitat for the endangered western pond turtle. Several local agencies are working to restore the park to its original riparian gallery forest conditions, including cottonwoods and willows. Today, although there are some mature cottonwoods and willows, the ground is covered with invasive reed canary grass, which sometimes grows so thickly that pond turtles are unable to build nests on the pond banks, and seedlings of native trees and other plants have a difficult time getting established. Another concern that's arisen is that reed canary grass grows so tall (it can grow to over 5 feet in a season) that it allows homeless people to hide their camps from patrolling police. This is leading to the development of a restoration plan that includes spraying the reed canary grass with herbicide for at least three years in order to establish the desired riparian vegetation and make the park less attractive to the homeless.

At one time, the area was home to numerous beavers that engineered their homes in such a way as to create slower moving waters. Braided networks of side channels filled the valley floors. The beavers were trapped nearly to local extinction between 1800 and 1950, drastically changing the nature of the watercourse. Then the Army Corps of Engineers built a dam to control flooding in the 1950s, further altering the course and character of the river. The artificially stabilized bank was subsequently mined for gravel to supply road construction throughout the area. Today, the dense rhizomatous growth of perennial reed canary grass stabilizes the bank. Over time, this encourages ponding and slows the velocity of water along its shores. The reed grass retains sediment near the headwaters of the watershed, keeping valuable soil resources upstream from the ocean.

It is true that if people want riparian trees to grow in the area, a patch of reed canary grass is not the best place for them, at least not within the next hundred years. Remember that this ecosystem is also on a successional pathway and that nothing, not even the most tenacious invasive species, will last forever, especially not a pioneering, early successional grass species like reed canary grass. If removing the reed canary grass is the best decision, there are many creative ways to do so that would take advantage of its potential yields while also steering the ecosystem toward native tree establishment.

A different approach would be for land managers to partner with the homeless population to construct simple, permanent dwellings. In exchange for the cost of materials and ongoing maintenance, residents could spend time each week cutting back the reed canary grass or keeping goats that graze rotationally on its lush growth. They could obtain yields by making goat cheese, growing a garden, keeping a native plant nursery, creating economic opportunities while stewarding the place where they live. Over time, with diligence and care, the park could become filled with native plants, and people could be gainfully employed and appropriately housed.

This approach would require communication, planning, training, and oversight, and, most importantly, a change in perception about what restoration as a practice should aim to accomplish. The current trajectory of this "nature park" is as a place where people pass through and don't stay to develop a relationship with it, making it all the more likely that invasive species will continue to thrive, even with standard herbicide-based management. Unless people are present, caring for the land as a part of their lives, planners are unlikely to reach their restoration goals and the "problem" of people camping in the park will continue.

Maybe what I'm describing is restoration at its most radical. There are so many opportunities, there is so much work to be done, and there are so many economic boons to be had from the creative integration of people with their landscapes. Granted, this vision doesn't fit well with conventional realities of land ownership, acquisition, and management, but these concepts merit consideration as we decide how to best engage in the process of good stewardship of land and the sustainable allocation of resources. The many problems of our modern society have common roots, and in addressing them with creative, integrated solutions, it is possible to imagine a future where the permaculture ethics of caring for Earth, caring for people, and reinvesting and redistributing surplus become commonplace.

Putting It All Together

Holistic restoration planning requires an honest accounting of what has come to pass as well as a comprehensive view of what we can do about it. The problems are complex, and the solutions are likely to be more so, at least at first. Navigating from a paradigm that views invasive species as scourges to one that looks at them as opportunities for deeper ecological and economic engagement will take time and commitment, especially because the old paradigm is so entrenched politically, economically, and academically. The tide is shifting though, as more and more of us are coming to realize that the herbicide-based eradication approach to restoration is outmoded – a futile attempt to regain an imagined past – and we need to be focusing our time, resources, and energy on adapting to the future.

To think about restoration in a larger sense means placing ourselves firmly within the ecosystems where we live and assisting in the enhancement of their biological diversity on multiple scales. Permaculture gives us a set of ethics and principles based on the patterns of natural systems that will allow us to more closely align our actions and decisions with the ecological processes that support our lives. We need to use these ethics, principles, and practices as we engage in the necessary work of the future: rebuilding, renewing, and restoring vital relationships of ecological stewardship.

Acknowledgments

I am deeply grateful for my family, friends, and community for their support and encouragement throughout this process. It has been a long labor of love. Thank you to Abel for washing the dishes for many months and for your support and critical feedback. Thanks to my mom for teaching me to ask the flowers before I picked them, because each one has a presence and a personality that I should respect first and foremost . . . you planted the seed for this book long ago. Thanks to Katie and Bill and my literary extended family for your encouragement, editing, love, and support. Thanks to Devon for offering your office space, as well as child care trades, and to Adrienne and Heather for hanging out with Sylvan while I wrote. Thank you to all of the yerba mate (*Ilex paraguariensis*) farmers of South America. I couldn't have done it without you. Thanks to all of my friends and colleagues in the permaculture, organic farming, and restoration fields for all of your hard work and dedication.

Acknowledgments

APPENDIX A

Species Identification List

The plants and animals identified in the book are listed by common names only for ease of reading. This list provides Latin names of those species for the purposes of proper identification.

Common Name	Species
Agave	*Agave* spp.
Ailanthus silkworm	*Bombyx cynthia*
Alder	*Alnus* spp.
Allegheny chinquapin	*Castanea pumila*
American chestnut	*Castanea dentata*
American ginseng	*Panax cinquefolium*
American sloughgrass	*Beckmannia syzigachne*
Arrowleaf balsamroot	*Balsamorhiza sagitatta*
Ash tree	*Fraxinus* spp.
Asian carp	*Cyprinus* spp.
Australian pine	*Casuarina* spp.
Australian tree fern	*Cyathea cooperi*
Barberry	*Mahonia haematocarpa*
Barred owl	*Strix varia*
Basin wildrye	*Elymus cinereus*
Beargrass	*Xerophyllum tenax*
Beech	*Fagus* spp.
Bergamot	*Monarda fistulosa*
Biscuitroot	*Lomatium* spp.
Bitterroot	*Lewisia rediviva*
Black cherry	*Prunus serotina*
Black cisco	*Coregonus nigripinnis*
Black-eyed Susan	*Rudbeckia hirta*
Black locust	*Robinia psuedoacacia*
Black nightshade	*Solanum americanum*
Black walnut	*Juglans nigra*
Black wattle	*Acacia mearnsii*
Bloodroot	*Sanguinaria canadensis*

Common Name	Species
Blue cohosh	*Caulophyllum thalictroides*
Boneset	*Eupatorium* spp.
Bradshaw's lomatium	*Lomatium bradshawii*
Breadfruit	*Artocarpus altilis*
Brown tree snake	*Boiga irregularis*
Buckbrush	*Ceanothus velutinus*
Bull thistle	*Cirsium vulgare*
Bunchberry	*Cornus canadensis*
Bush honeysuckle	*Lonicera japonica*
California yellow lupine	*Lupinus arboreus*
Camas	*Camassia quamash*
Canada thistle	*Cirsium arvense*
Candleberry	*Myrica faya*
Cane toad	*Rhinella marina*
Cascara sagrada	*Rhamnus purshiana*
Cat's claw	*Uncaria tomentosa*
Ceanothus	*Ceanothus* spp.
Cheat grass	*Bromus tectorum*
Checker mallow	*Sidalcea* spp.
Chia	*Salvia hispanica*
Common buckthorn	*Rhamnus cathartica*
Common reed	*Phragmites australis*
Corn	*Zea mays*
Crabapple	*Malus* spp.
Dandelion	*Taraxacum officinale*
Dawn redwood	*Metasequoia glyptostroboides*

Common Name	Species
Dwarf trout lily	*Erythronium propullans*
Echinacea	*Echinacea* spp.
Eelgrass	*Zostera japonica*
Elderberry	*Sambucus* spp.
Emerald ash borer	*Agrilus planipennis*
English ivy	*Hedera helix*
Ephedra	*Ephedra* spp.
Eurasian water milfoil	*Myriophyllum spicatum*
False dandelion	*Hypocharis radicata*
Field bindweed	*Convolvulus arvensis*
Fleabane	*Erigeron* spp.
Florida torreya	*Torreya taxifolia*
Fremont cottonwood	*Populus fremontii*
Garlic mustard	*Alliaria petiolata*
Geranium filaree	*Erodium* spp.
Ghost shrimp	*Neotrypaea californiensis*
Giant goldenrod	*Solidago gigantea*
Giant hogweed	*Heracleum mantegazzianum*
Giant reed	*Arundo donax*
Gingko	*Gingko biloba*
Glossy buckthorn	*Rhamnus frangula*
Goldenseal	*Hydrastis canadensis*
Goodding's willow	*Salix gooddingii*
Gooseberry	*Ribes uva-crispa*
Grain sorghum	*Sorghum bicolor*
Guam flycatcher	*Myiagra freycineti*
Guava	*Psidium guajava*
Hackberry	*Celtis* spp.
Guam rail	*Gallirallus owstoni*
Harvest lily	*Dichelostemma congesta*
Hawaiian myrtle	*Metrosideros polymorpha*
Hawthorn	*Crataegus* spp.
Hickory	*Carya* spp.
Highbush cranberry	*Viburnum trilobum*
Himalayan blackberry	*Rubus armeniacus*
Hoary cress	*Lepidium draba*
Honey locust	*Gleditsia triacanthos*
Hornbeam	*Carpinus caroliniana*
Huckleberry	*Vaccinium* spp.
Hydrilla	*Hydrilla verticillata*
Japanese knotweed	*Polygonum cuspidatum*
Johnson grass	*Sorghum halapense*

Common Name	Species
Jojoba	*Simmondsia chinensis*
King bolete	*Boletus edulis*
Koa	*Acacia koaia*
Kudzu	*Pueraria lobata*
Lama	*Diospyros sandwicensis*
Leafy spurge	*Euphorbia esula*
Lemonade berry	*Rhus integrifolia*
Linden	*Tilia* spp.
Little button mushroom	*Agaricus bisporus*
Lobelia	*Lobelia* spp.
Longleaf pine	*Pinus palustris*
Mahogany	*Swietenia* spp.
Mallow	*Malva* spp.
Mesquite	*Prosopis* spp.
Mile-a-minute vine	*Persicaria perfoliata*
Milkweed	*Asclepias* spp.
Monterey pine	*Pinus radiata*
Mother-of-millions	*Bryophyllum* spp.
Mountain mahogany	*Cercocarpus traskiae*
Mud shrimp	*Upogebia pugettensis*
Mulberry	*Morus* spp.
Mulberry silkworm	*Bombyx mori*
Mule's ears	*Wyethia angustifolia*
Multiflora rose	*Rosa multiflora*
Musk thistle	*Carduus nutans*
Myrtle	*Myrica* spp.
Nannyberry	*Viburnum lentago*
Narrowleaf onion	*Allium amplectens*
Nasturtium	*Tropaeolum majus*
Nile perch	*Lates niloticus*
Nodding onion	*Allium cernuum*
Northern spotted owl	*Strix occidentalis*
Oak	*Quercus* spp.
Ocotillo	*Fouquieria splendens*
Oregon gumweed	*Grindelia stricta*
Palmer amaranth	*Amaranthus palmeri*
Palo verde	*Parkinsonia microphylla*
Pampas grass	*Cortaderia selloana*
Pawpaw	*Asimina triloba*
Peach palm	*Bactris gasipaes*
Pennyroyal	*Mentha pulegium*
Perennial chiles	*Capsicum* spp.

Species Identification List

Common Name	Species
Persimmon	*Diospyros* spp.
Pinyon pine	*Pinus edulis*
Prickly pear	*Opuntia* spp.
Privet	*Ligustrum obtusifolium*
Purple loosestrife	*Lythrum salicaria*
Rattail fescue	*Vulpia myuros*
Red maple	*Acer rubrum*
Red mulberry	*Morus rubra*
Red stem dogwood	*Cornus sericea*
Redbud	*Cercis occidentalis*
Reed canary grass	*Phalaris arundinaceae*
Russian thistle	*Salsola* spp.
Sagebrush	*Artemisia* spp.
Salt cedar	*Tamarisk* spp.
Sassafras	*Sassafras* spp.
Saw palmetto	*Serenoa repens*
Scotch broom	*Cystisus scoparius*
Sea buckthorn	*Hippophae rhamnoides*
Sedge	*Carex* spp.
Self-heal	*Prunella vulgaris*
Serviceberry	*Amelanchier* spp.
Silk tree	*Albizia julibrissin*
Silverberry	*Eleagnus* spp.
Sourwood	*Oxydendrum arboreum*
Sow thistle	*Sonchus* spp.
Soy	*Glycine max*
Spartina	*Spartina alterniflora*
Spiraea	*Spiraea* spp.
Spotted frog	*Rana pretiosa*
Spotted knapweed	*Centaurea maculosa*
St. John's wort	*Hypericum perforatum*
Strawberry	*Fragaria* spp.

Common Name	Species
Sugar maple	*Acer saccharum*
Sweet potato	*Ipomoea batatas*
Tangan-tangan	*Leucaena leucocephala*
Tarweed	*Madia* spp.
Teak	*Tectona grandis*
Thistle head weevil	*Rhinocyllus conicus*
Tree of heaven	*Ailanthus altissima*
Trillium	*Trillium* spp.
Tufted hairgrass	*Deschampsia cespitosa*
Tule	*Schoenoplectus acutus*
Umari	*Poraqueiba sericea*
Wapato	*Sagittaria latifolia*
Water hyacinth	*Eichornia cressipes*
Water lily	*Nuphar polysepala*
Water plantain	*Alisma plantago-aquatica*
Western pond turtle	*Actinemys marmorata*
Western willow flycatcher	*Empidonax traillii*
Wheat	*Triticum* spp.
Wild ginger	*Asarum canadense*
Wild rice	*Zizania palustris*
Wiliwili	*Erythrina sandwicensis*
Willow herb	*Epilobium* spp.
Windmill palm	*Trachycarpus fortunei*
Wintergreen	*Gaultheria procumbens*
Wolfberry	*Lycium* spp.
Yampah	*Perideridia gairdneri*
Yellow foxtail	*Setaria pumila*
Yellow nut sedge	*Cyperus edulis*
Yellow star thistle	*Centaurea solstitialis*
Yucca	*Yucca* spp.
Zebra mussel	*Dreisennia polymorpha*

APPENDIX B

Organisms that Moved around the World Prior to the Modern Colonial Era

TABLE B.1. Plants for which there is decisive evidence of transoceanic movement

Species	Common Name	Origin	Moved To
Adenostemma viscosum	dungweed	Americas	Hawaii
Agave americana	agave	Americas	India
Agave angustifolia	agave	Americas	India
Agave cantala	agave	Americas	India
Ageratum conyzoides	goat weed	Americas	India, Marquesas
Alternanthera spp.	alligator weed	Americas	India
Amaranthus caudatus	love-lies-bleeding	Americas	Asia
Amaranthus cruentus	amaranth	Americas	Asia
A. hypochondriacus	amaranth	Americas	Asia
Amaranthus spinosus	spiked amaranth	Americas	South Asia
Anacardium occidentale	cashew	Americas	India
Ananas comosus	pineapple	Americas	India, Polynesia
Annona cherimola	large Annona	Americas	India
Annona reticulata	custard apple	Americas	India
Annona squamosa	sweetsop	Americas	India, Timor
Arachis hypogaea	peanut	Americas	China, India
Argemone mexicana	Mexican poppy	Americas	China, India
Aristida subspicata	three-awn grass	Americas	Polynesia
Artemisia vulgaris	mugwort	Eastern Hemisphere	Mexico
Asclepias curassavica	milkweed	Americas	China, India
Aster divaricatus	white wood aster	Americas	Hawaii
Bixa orellana	achiote, annatto	Americas	Oceania, Asia
Canavalia sp.	sword bean	Americas	Asia
Canna edulis	achira	Americas	India, China
Cannabis sativa	marijuana	Eastern Hemisphere	Peru
Capsicum annuum	chili pepper	Americas	India, Polynesia

Species	Common Name	Origin	Moved To
Capsicum frutescens	chili pepper	Americas	India
Carica papaya	papaya	Americas	Polynesia
Ceiba pentandra	kapok	Americas	Asia
Chenopodium ambrosioides	epazote	Eastern Hemisphere	Mesoamerica
Cocos nucifera	coconut	Eastern Hemisphere	Colombia to Mexico
Couroupita guianensis	cannonball tree	Americas	India
Cucurbita ficifolia	chilacayote	Americas	Asia
Cucurbita maxima	Hubbard squash	Americas	India, China
Cucurbita moschata	butternut squash	Americas	India, China
Cucurbita pepo	pumpkin	Americas	India, China
Curcuma longa	turmeric	Eastern Hemisphere	Andes
Cyperus esculentus	yellow nut sedge	Americas	Eurasia
Cyperus vegetus	edible sedge	Americas	India, Easter Island
Datura metel	jimsonweed	Americas	Eurasia
Datura stramonium	thorn apple	Americas	Eurasia
Diospyros ebenaster	black sapote	Americas	Eurasia
Erigeron canadensis	Canadian fleabane	Americas	India
Erythroxylum novagranatense	coca	Americas	Egypt
Garcinia mangostana	mangosteen	Eastern Hemisphere	Peru
Gossypium arboreum	a cotton	Eastern Hemisphere	South America
Gossypium barbadense	a cotton	Americas	Marquesas Islands
Gossypium gossypioides	a cotton	Africa	Mexico
Gossypium hirsutum	a cotton	Mexico	Africa, Polynesia
Gossypium tomentosum	a cotton	Americas	Hawaii
Helianthus annuus	sunflower	Americas	India
Heliconia bihai	macaw flower	Americas	Oceania, Asia
Hibiscus tiliaceus	sea hibiscus	Americas	Polynesia
Ipomoea batatas	sweet potato	Americas	Polynesia, China
Lagenaria siceraria	bottle gourd	Americas	Asia, East Polynesia
Luffa acutangula	ribbed gourd	Americas	India
Luffa cylindrica	loofa	Americas	India, China
Lycium carolinianum	creeping wolfberry	Americas	Easter Island
Macroptilium lathyroides	phasey bean	Americas	India
Manihot sp.	manioc	Americas	Easter Island
Maranta arundinacea	arrowroot	Americas	Easter Island, India
Mimosa pudica	sensitive plant	Americas	India, China

Organisms that Moved around the World

Species	Common Name	Origin	Moved To
Mirabilis jalapa	four-o'clock	Americas	India
Mollugo verticillata	carpetweed	Eastern Hemisphere	North America
Monstera deliciosa	Mexican breadfruit	Americas	India
Morus sp.	mulberry	Eastern Hemisphere	Middle America
Mucuna pruriens	cowhage	Americas	India, Hawaii
Musa × paradisiaca	banana, plantain	Eastern Hemisphere	Tropical America
Myrica gale	bog myrtle	Eastern Hemisphere	North America
Nicotiana tabacum	tobacco	Americas	South Asia
Ocimum sp.	basil	Americas	India
Opuntia dillenii	prickly pear cactus	Americas	India
Osteomeles anthyllidifolia	Hawaiian hawthorn	Americas	China, Oceania
Pachyrhizus erosus	jicama	Americas	Asia
Pachyrhizus tuberosus	jicama, yam bean	Americas	India, China, Oceania
Pharbitis hederacea	ivyleaf morning glory	Americas	India, China
Phaseolus lunatus	lima bean	Americas	India
Phaseolus vulgaris	kidney bean	Americas	India
Physalis sp.	ground cherry	Americas	China
Physalis peruviana	husk tomato	Americas	East Polynesia
Polygonum acuminatum	a knotweed	Americas	Easter Island
Portulaca oleracea	purslane	Americas	Eurasia
Psidium guajava	guava	Americas	China, Polynesia
Salvia coccinea	scarlet salvia	Americas	Marquesas Islands
Sapindus saponaria	soapberry	Americas	India, East Polynesia
Schoenoplectus californicus	bulrush, totora reed	Americas	Easter Island
Sisyrinchium acre	Hawaii blue-eyed grass	Americas	Hawaii
Sisyrinchium angustifolium	blue-eyed grass	Americas	Greenland
Smilax sp.	sarsparilla	Eastern Hemisphere	Central America
Solanum candidum	naranjillo	Americas	Oceania, Southeast Asia
Solanum nigrum	black nightshade	Eastern Hemisphere	Mesoamerica
Solanum repandum	cocona	Americas	Oceania
Solanum tuberosum	potato	Americas	Easter Island
Sonchus oleraceus	sow thistle	Americas	China
Sophora toromiro	toromiro tree	Americas	Easter Island
Tagetes erecta	marigold	Americas	India, China
Tagetes patula	dwarf marigold	Americas	India, Persia
Zea mays	corn, maize	Americas	Eurasia, Africa

TABLE B.2. **Plants for which evidence is significant but not decisive**

Species	Common Name	Origin	Moved To
Acorus calamus	sweet flag	Asia	North America
Amanita muscaria	fly agaric	Europe	Middle America
Chenopodium quinoa	quinoa	Easter Island	South America
Erigeron albidus	Argentine fleabane	Hawaii	South America
Gnaphalium purpureum	Jersey cudweed	Hawaii	South America
Indigofera sp.	indigo	Polynesia, Asia	Mexico, Peru
Ipomoea acetosaefolia	beach morning glory	Hawaii	South America
Mangifera indica	mango	Asia	Central America
Musa coccinea	Chinese banana	China	South America
Nelumbo nucifera	East Indian lotus	East Indies, India	Tropics
Nymphaea nouchali	Indian water lily	India, Egypt	Tropics
Phaseolus adenanthus	manding-bambara	Polynesia	South America
Saccharum officinarum	sugarcane	Asia	South America
Salvia occidentalis	a sage	Marquesas	South America
Triumfetta semitriloba	burweed	Easter Island	South America
Verbesina encelioides	cowpen daisy	Hawaii	South America
Vitis vinifera	grape	Eurasia	Mexico

TABLE B.3. Plants for which evidence justifies further study

Species	Common Name
Ageratum houstonianum	floss flower
Annona glabra	pond apple
Cajanus cajan	pigeon pea
Cassia fistula	golden shower tree
Cinchona officinalis	quinine (bark)
Colocasia esculenta	dry-land taro
Cucumis sp.	cucumber and melons
Cyclanthera pedata	pepino hueco
Datura sanguinea	red angel's trumpet
Derris sp.	timborana
Dioscorea alata	yam
Dioscorea cayenensis	guinea yam
Dolichos lablab	lablab bean
Elaeis guineensis	guinea oil palm
Gossypium brasiliense	a cotton
Gossypium drynarioides	a cotton
Gossypium religiosum	a cotton
Hibiscus youngianus	linden-leaf hibiscus
Indigofera tinctoria	indigo
Lonchocarpus sericeus	Senegal lilac
Lupinus cruickshanksii	field lupine
Lycopersicon esculentum	tomato
Nicotiana rustica	wild tobacco
Ocimum americanum	hoary basil
Paullinia spp.	soapberry
Sagittaria sagittifolia	wapato
Sesamum orientale	sesame
Spondias lutea	yellow mombin
Spondias purpurea	hog plum
Synedrella nodiflora	Cinderella weed
Tamarindus indicus	tamarind tree
Tephrosia spp.	fish-poison bean
Trapa natans	water chestnut
Vigna sinensis	cowpea

TABLE B.4. Animals for which there is decisive evidence of pre-Columbian exchange

Species	Common Name
Alphitobius diaperinus	lesser mealworm
Gallus gallus	chicken
Littorina littorea	a mollusk
Meleagris gallopavo	turkey
Mya arenaria	American soft-shell clam
Stegobium paniceum	drugstore beetle

TABLE B.5. Other animals needing additional study

Species	Common Name
Cairina moschata	Muscovy duck
Canis familiaris	dog
Crax globicera	curassow
Cicada spp.	cicada
Dendrocygna bicolor	fulvous tree duck
Lasioderma serricorne	tobacco, or cigarette, beetle
Rhyzopertha dominica	lesser grain borer

Source: Adapted from John L. Sorenson and Carl Johannessen, "Scientific Evidence for Pre-Columbian Transoceanic Voyages to and from the Americas," accessed June 4, 2014, http://publications.maxwellinstitute.byu.edu/fullscreen/?pub=1068&index=1.

APPENDIX C

Recommended Resources for Holistic Restoration and Invasive Species Management Alternatives

Books

Anderson, Kat. *Tending the Wild: Native American Knowledge and the Management of California's Natural Resources*. Berkeley: University of California Press, 2005.

Berkes, Fikret. *Sacred Ecology: Traditional Ecological Knowledge and Resource Management*. Oxford, UK: Taylor & Francis, 1999.

Deur, Douglas, and Nancy J. Turner, eds. *Keeping It Living: Traditions of Plant Use and Cultivation on the Northwest Coast of North America*. Seattle: University of Washington Press, 2005.

Gunderson, Lance H., C. S. Holling, and Stephen S. Light. *Barriers and Bridges to the Renewal of Ecosystems and Institutions*. New York: Columbia University Press, 1995.

Holmgren, David. *Principles & Pathways Beyond Sustainability*. Hepburn, Australia: Holmgren Design Services, 2002.

Holmgren, David. *Trees on the Treeless Plains*. Hepburn, Australia: Holmgren Design Services, 1994.

Jacke, David, with Eric Toensmeier. *Edible Forest Gardens*. White River Junction, VT: Chelsea Green Publishing, 2005.

Jackson, Wes. *New Roots for Agriculture*. Lincoln: University of Nebraska Press, 1980.

Kuhn, Thomas S. *The Structure of Scientific Revolutions*. Chicago: University of Chicago Press, 1962.

Meadows, Donella H. *Thinking in Systems: A Primer*. White River Junction, VT: Chelsea Green Publishing, 2008.

Margulis, Lynn. *Symbiotic Planet: A New Look at Evolution*. New York: Basic Books, 2008.

Mollison, Bill. *Permaculture: A Designer's Manual*. Tyalgum, Australia: Tagari Press, 1988.

Pilarski, M., ed. *Restoration Forestry: An International Guide to Sustainable Forestry Practices*. Skyland, NC: Kivaki Press, 1994.

Ross, Anne, Kathleen Pickering Sherman, Jeffrey G. Snodgrass, Henry D. Delcore, and Richard Sherman. *Indigenous Peoples and the Collaborative Stewardship of Nature: Knowledge Binds and Institutional Conflicts*. Walnut Creek, CA: Left Coast Press, 2011.

Savory, Allan, and Jody Butterfield. *Holistic Management: A New Framework for Decision Making*. Washington, DC: Island Press, 1999.

Stewart, Omer Call. *Forgotten Fires: Native Americans and the Transient Wilderness*. Norman: University of Oklahoma Press, 2002.

Sorenson, John L., and Carl L. Johannessen. *World Trade and Biological Exchanges Before 1492*. Bloomington, IN: iUniverse, 2009.

Turner, Nancy J. *The Earth's Blanket*. Vancouver, Canada: Douglas & McIntyre, 2005.

Walker, Brian, and David Salt. *Resilience Thinking: Sustaining Ecosystems and People in a Changing World*. Washington, DC: Island Press, 2006.

Yeomans, Percival Alfred. *The City Forest: The Keyline Plan for the Human Environment Revolution*. Sydney, Australia: Keyline Press, 1971.

Zeedyk, Bill, and Van Clothier. *Let the Water Do the Work: Induced Meandering, an Evolving Method for Restoring Incised Channels*. White River Junction, VT: Chelsea Green Publishing, 2014.

Websites

Aprovecho
http://www.aprovecho.net

Beyond Pesticides
http://www.beyondpesticides.org

BONAP – Biota of North America Program
http://www.bonap.org

Eat the Invaders: Fighting Invasive Species – One Bite at a Time
http://www.eattheinvaders.org

Ewe4ic Ecological Services – Lani Malmberg
http://www.goatseatweeds.com

Forager's Harvest
http://www.foragersharvest.com

Holistic Management
http://www.holisticmanagement.org
http://www.savoryinstitute.org

Indigenous People's Restoration Network (IPRN)
http://www.ser.org/iprn

Invasivore
http://www.invasivore.org

J. L. Hudson, Seedsman
http://www.jlhudsonseeds.net

Lomakatsi Restoration Project
http://www.lomakatsi.org

Mattole Restoration Council
http://www.mattole.org

Natural Resource Conservation Service (NRCS) Indigenous Stewardship Methods
http://www.nrcs.usda.gov/wps/portal/nrcs/detail/national/plantsanimals/plants/?cid=stelprdb1045246

Northwest Coalition for Alternatives to Pesticides
http://www.pesticide.org

Oliver Kellhammer
http://www.oliverk.org

Permaculture Institute of North America (PINA)
http://www.pina.in

Philippe Parola – Chef
http://www.chefphilippe.com

Plants for a Future
http://www.pfaf.org

Quivira Coalition
http://www.quiviracoalition.org

Resilience Permaculture Design
http://www.resiliencepermaculture.com

Stockholm Resilience Centre
http://www.stockholmresilience.org

The Resilience Alliance
http://www.resalliance.org

Torreya Guardians
http://www.torreyaguardians.org

Weedrobes
http://www.nicoledextras.com

Notes

Introduction

1. David S. Wilcove, David Rothstein, Jason Dubow, Ali Phillips, and Elizabeth Losos, "Quantifying threats to imperiled species in the United States," *BioScience* (1998): 607–615.
2. International Union for Conservation of Nature (IUCN), "The Fight Against Invasives," accessed January 19, 2015, http://www.iucn.org/news_homepage/news_by_date/?7634/The-Fight-Against-Invasives.
3. National Audubon Society, "How Invasive Species Threaten Habitat," accessed January 26, 2015, http://www.massaudubon.org/our-conservation-work/ecological-management/invasive-species, and National Wildlife Federation, "Invasive Species," accessed January 26, 2015, http://www.nwf.org/Wildlife/Threats-to-Wildlife/Invasive-Species.aspx.
4. Bureau of Land Management (BLM), "BLM's War Against Weeds in Nevada," accessed January 26, 2015, http://www.blm.gov/nv/st/en/prog/more_programs/invasive_species.print.html, and US Forest Service, "Four Threats to the Health of the Nation's Forests and Grasslands," accessed January 26, 2015, http://www.fs.fed.us/projects/four-threats/#species.
5. The Nature Conservancy, "Invasive Species," accessed June 12, 2014, http://www.nature.org/ourinitiatives/regions/northamerica/unitedstates/southcarolina/placesweprotect/invasives-fact-sheet-final.pdf.
6. David Pimentel, S. McNair, J. Janecka, J. Wightman, C. Simmonds, C. O'Connell, E. Wong, L. Russel, J. Zern, T. Aquino, and T. Tsomondo, "Economic and Environmental Threats of Alien Plant, Animal, and Microbe Invasions," *Agriculture, Ecosystems & Environment* 84, no. 1 (2001): 1–20.
7. National Invasive Species Council, "Meeting the Invasive Species Challenge," (January, 2001): 1, http://www.invasivespeciesinfo.gov/docs/council/mp.pdf.
8. Bill Mollison, "Permaculture Techniques" Pamphlet IX, Permaculture Design Course Series, (1981), 13, http://www.barkingfrogspermaculture.org/PDC_ALL.pdf.

Chapter 1: Against All Ethics

1. Charles S. Elton, *The Ecology of Invasions by Animals and Plants* (Chicago: University of Chicago Press, 2000), 18.
2. Mark A. Davis, Ken Thompson, and J. Philip Grime, "Charles S. Elton and the Dissociation of Invasion Ecology From the Rest of Ecology," *Diversity and Distributions* 7, no. 1–2 (2001): 97–102.
3. Elton, *The Ecology of Invasions*,15.
4. John H. Perkins, "The Rockefeller Foundation and the Green Revolution, 1941–1956," *Agriculture and Human Values* 7, no. 3–4 (1990): 6–18.
5. William A. Buckingham Jr., *Operation Ranch Hand: The Air Force and Herbicides in Southeast Asia, 1961–1971*, Office of Air Force History, Washington DC, (1982).
6. Irwin N. Forseth Jr. and Anne F. Innis, "Kudzu (*Pueraria montana*): History, Physiology, and Ecology Combine to Make a Major Ecosystem Threat," *Critical Reviews in Plant Sciences*, 23 (2004): 401–413.
7. US Congress Office of Technology Assessment, *Harmful Non-Indigenous Species in the United States*, US Government Printing Office (1993): 3.

8. National Invasive Species Council, "Five Year Review of Executive Order 13112 on Invasive Species," 2005, http://www.invasivespeciesinfo.gov/docs/council/fiveyearreview.pdf.
9. Invasive Species Advisory Committee, "Invasive Species Advisory Committee Meeting: Summary," National Invasive Species Council, 2005, http://www.invasivespecies.gov/global/ISAC/ISAC_Minutes/2005/ISAC%20Minutes%20October%2011-13%202005%20Washington.pdf.
10. Definitions Subcommittee of the Invasive Species Advisory Committee (ISAC), "Invasive Species Definition Clarification and Guidance White Paper," National Invasive Species Council, April 27, 2006, http://www.invasivespeciesinfo.gov/docs/council/isacdef.pdf.
11. Cal-IPC, "2008 Cal-IPC Symposium Sponsors," 2008, http://www.cal-ipc.org/symposia/pdf/2008%20Cal-IPC%20Symposium%20Sponsors.pdf.
12. Invasive Species Advisory Committee, "Member Bios," National Invasive Species Council, accessed January 14, 2015, http://www.hear.org/hwgpisac/isacmemberbios.pdf.
13. Nelson E. Jackson, "Control of *Arundo donax*: Techniques and Pilot Projects," presented at the *Arundo donax* Workshop Proceedings, 1993, http://www.cal-ipc.org/symposia/archive/pdf/Arundo_Proceedings_1993.pdf.
14. Invasive Species Advisory Committee, "Members 2002–2004," National Invasive Species Council, accessed November 7, 2014, http://www.invasivespecies.gov.
15. Weed Science Society of America, "2011 Annual Meeting Titles and Abstracts," 2011, accessed March 25, 2014, http://wssaabstracts.com/public/4/Presentations-sorted-by-title.html.
16. Bureau of Land Management (BLM), "Herbicides Currently Approved and Herbicides to be Examined for use on BLM Lands in Oregon," July 2008, http://www.blm.gov/or/plans/vegtreatmentseis/files/Herbicides_chart.pdf.
17. Food and Water Watch, "Public Research, Private Gain," April 2012, http://documents.foodandwaterwatch.org/doc/PublicResearchPrivateGain.pdf.
18. New Mexico State University College of Agricultural, Consumer, and Environmental Sciences, "Brush and Weed Control Education Program (Lincoln County)," last updated November 12, 2009, accessed January 14, 2015, http://pow.nmsu.edu/view_plan.php?plan_id=349&page=1.
19. Linda Lear, *Rachel Carson: Witness for Nature* (New York: Macmillan, 1998), 477.
20. The Nature Conservancy, "Working with Companies: Companies We Work With," accessed December 10, 2013, http://www.nature.org/about-us/working-with-companies/companies-we-work-with/index.htm.
21. The Nature Conservancy, "Global Invasive Species Team Listserv Digest #053," January 25, 2000, accessed January 15, 2015, http://www.invasive.org/gist/listarch/arch053.html.
22. Catherine Cluett, "Monsanto Fund Donates $20,000 for Molokai Watershed Protection," *Molokai Dispatch*, October 14, 2012.
23. National Audubon Society. "Waterbirds on Working Lands," accessed February 18, 2015, http://birds.audubonstg.zivtech.com/waterbirds-working-lands-project.
24. Ducks Unlimited Canada, "Annual Report 2012: Our Conservation Supporters," June 2012, http://www.ducks.ca/assets/2012/06/duc-annualreport2012-supporters.pdf.
25. Southeastern Louisiana University, "Southeastern's Turtle Cove Receives Grant from Monsanto," news release, August 2010, accessed February 25, 2014, http://www.southeastern.edu/news_media/news_releases/2010/aug/monsanto_fund.html.
26. BLM, "Learning Landscapes: Invasive Species," last updated September 1, 2011, accessed February 25, 2014, http://www.blm.gov/wo/st/en/res/Education_in_BLM/Learning_Landscapes/For_Kids/homework_helpers/invasive_species.html.
27. United States Environmental Protection Agency (US EPA), "Pesticides Industry Sales and Usage: 2006 and 2007 Market Estimates," www.epa.gov/opp00001/pestsales/07pestsales/market_estimates2007.pdf, 12.

28. US EPA, "Explanation of Statutory Framework for Risk-Benefit Balancing for Public Health Pesticides," last updated May 9, 2012, accessed June 14, 2104, http://epa.gov/pesticides/health/risk-benefit.htm.
29. US EPA, "Consumer Factsheet on: Atrazine," accessed June 14, 2014, http://www.epa.gov/safewater/pdfs/factsheets/soc/atrazine.pdf.
30. US Government Printing Office, "Electronic Code of Federal Regulations," accessed March 14, 2014, http://www.ecfr.gov/cgi-bin/text-idx?c=ecfr&sid=bd32aab1f2263d189c2ea7ae45c321e9&tpl=/ecfrbrowse/Title40/40cfr180_main_02.tpl.
31. US EPA Office of Pesticide Programs, "Index of Chemical Names, Part 180 Tolerance Information, and Food and Feed Commodities (by Commodity)," December 12, 2012, http://www.epa.gov/pesticides/regulating/tolerances-commodity.pdf.
32. US EPA, "Common Mechanism Groups; Cumulative Exposure and Risk Assessment," last updated May 9, 2012, accessed January 18, 2015, http://www.epa.gov/pesticides/cumulative/common_mech_groups.htm.
33. US Government Accountability Office, "Federal Research: NIH and EPA Need to Improve Conflict of Interest Reviews for Research Arrangement with Private Sector Entities," February 2005, http://www.iecjournal.org/iec/files/gao_report05191.NIH%20and%20EPA%20Need%20to%20Improve%20Conflict%20of%20Interest%20REviews%20for%20Research%20Arrangements%20wtin%20Private%20Sector%20Entities.pdf.
34. Jennifer Sass and Mae Wu, "Superficial Safeguards: Most Pesticides Are Approved by Flawed EPA Process," *NRDC Issue Brief*, March 2013, http://www.nrdc.org/health/pesticides/files/flawed-epa-approval-process-IB.pdf, 2.
35. Monsanto Company, "Safety Data Sheet: Roundup Pro Herbicide," http://www.cdms.net/LDat/mp07A005.pdf, 5.
36. Leah Schinasi and Maria E. Leon, "Non-Hodgkin Lymphoma and Occupational Exposure to Agricultural Pesticide Chemical Groups and Active Ingredients: A Systematic Review and Meta-Analysis," *International Journal of Environmental Research and Public Health* 11, no. 4 (2014): 4449–4527.
37. http://www.epa.gov/espp/consultation/ecorisk-overview.pdf, 32.
38. Meriel Watts, "Glyphosate," Pesticide Action Network Aotearoa New Zealand, November 2009, http://www.pananz.net/wp-content/uploads/2013/04/Glyphosate-monograph-PANANZ.pdf, 24.
39. US EPA, "OCSPP Harmonized Test Guidelines – Master List," last updated February 2013, http://www.epa.gov/ocspp/pdfs/OCSPP-TestGuidelines_MasterList.pdf.
40. US EPA, "2006–2007 Pesticide Market Estimates: Usage," last updated July 9, 2013, accessed February 15, 2014, http://www.epa.gov/opp00001/pestsales/07pestsales/usage2007.htm.
41. Markets and Markets, "Agricultural Adjuvants Market by Type (Activator and Utility), Application (Herbicides, Fungicides, and Insecticides), & Region: Global Trends and Forecast to 2019," January 2015, accessed January 2, 2105, http://www.marketsandmarkets.com/Market-Reports/adjuvant-market-1240.html.
42. Monsanto, "Where Weeds and Water Meet," February 2008, http://www.monsanto.com/sitecollectiondocuments/aquamaster-ito.pdf.
43. Mandy Tu, Callie Hurd, and John M. Randall, "Weed Control Methods Handbook: Tools and Techniques for Use in Natural Areas," 2001, http://www.invasive.org/gist/products/handbook/07.herbicideguidelines.pdf, 8.6.
44. Ibid., 8.4.
45. US Government Publishing Office, "Code of Federal Regulations: 153.125 – Criteria for Determination of Pesticidal Activity," 2011, accessed January 18, 2015, http://www.gpo.gov/fdsys/pkg/CFR-2011-title40-vol24/pdf/CFR-2011-title40-vol24-sec153-125.pdf.
46. M. H. Surgan and A. G. Gershon, *The Secret Ingredients in Pesticides: Reducing the Risk* (New York: Office of the Attorney General Environmental Protection Bureau: 2000).

47. Holly Knight and Caroline Cox, "Worst Kept Secrets: Toxic Inert Ingredients in Pesticides," Northwest Coalition for Alternatives to Pesticides, January 1998, http://nctc.fws.gov/resources/course-resources/pesticides/Limitations%20and%20Uncertainty/ActiveInertsRpt.pdf.
48. Caroline Cox and Michael Surgan, "Unidentified Inert Ingredients in Pesticides: Implications for Human and Environmental Health," *Environmental Health Perspectives* (2006): 1803–1806.
49. Adriano Martínez, Ismael Reyes, and Niradiz Reyes, "Cytotoxicity of the Herbicide Glyphosate in Human Peripheral Blood Mononuclear Cells," *Biomedica: Revista del Instituto Nacional de Salud* 27, no. 4 (2007): 594–604.
50. Sophie Richard, Safa Moslemi, Herbert Sipahutar, Nora Benachour, and Gilles-Eric Seralini, "Differential Effects of Glyphosate and Roundup on Human Placental Cells and Aromatase," *Environmental Health Perspectives* 113, no. 6 (2005): 716.
51. Sally M. Bradberry, Alex T. Proudfoot, and J. Allister Vale, "Glyphosate Poisoning," *Toxicological Reviews* 23, no. 3 (2004): 159–167.
52. Nora Benachour and Gilles-Eric Seralini, "Glyphosate Formulations Induce Apoptosis and Necrosis in Human Umbilical, Embryonic, and Placental Cells," *Chemical Research in Toxicology* 22, no. 1 (2008): 97–105.
53. Markets and Markets, "Agricultural Adjuvants Market by Type," 6.14.
54. Michio Umegaki, Vu Le Thao Chi, and Tran Duc Phan, "Embracing Human Insecurity: Agent Orange-Dioxin and the Legacies of the War in Viet Nam," In *Human Insecurity in East Asia*, ed. Michio Umegaki, Lynn Thiesmeyer, and Atsushi Watabe (Tokyo: United Nations University Press, 2009), 21.
55. US EPA, "2,4-D RED Facts," accessed June 14, 2014, http://www.epa.gov/oppsrrd1/REDs/factsheets/24d_fs.htm.
56. J. A. Gervais, B. Luukinen, K. Buhl, D. Stone, "2,4-D Technical Fact Sheet," National Pesticide Information Center, Oregon State University Extension Services, 2008, http://npic.orst.edu/factsheets/2,4-DTech.pdf.
57. Helen H. McDuffie, Punam Pahwa, John R. McLaughlin, John J. Spinelli, Shirley Fincham, James A. Dosman, Diane Robson, Leo F. Skinnider, and Norman W. Choi. "Non-Hodgkin's Lymphoma and Specific Pesticide Exposures in Men Cross-Canada Study of Pesticides and Health," *Cancer Epidemiology Biomarkers & Prevention* 10, no. 11 (2001): 1155–1163.
58. Larry W. Figgs, Nina Titenko Holland, Nathanial Rothman, Shelia Hoar Zahm, Robert E. Tarone, Robert Hill, Robert F. Vogt, Martyn T. Smith, Cathy D. Boysen, Frederick R. Holmes, Karen VanDyck, and Aaron Blair, "Increased Lymphocyte Replicative Index Following 2, 4–Dichlorophenoxyacetic Acid Herbicide Exposure," *Cancer Causes & Control* 11, no. 4 (2000): 373–380.
59. US EPA, "Risks of 2,4-D Use to the Federally Threatened California Red-legged Frog (*Rana aurora draytonii*) and Alameda Whipsnake (*Masticophis lateralis euryxanthus*): Appendix E: Review of Dioxin Contamination," February 20, 2009, http://www.epa.gov/espp/litstatus/effects/redleg-frog/2-4-d/appendix-e.pdf.
60. US EPA, "Memorandum: Effects Determination for 2,4-D Relative to the California Red-Legged Frog and the Alameda Whipsnake and their Designated Critical Habitats," February 20, 2009, http://www.epa.gov/espp/litstatus/effects/redleg-frog/2-4-d/transmittal-memo.pdf.
61. Caroline Cox, "Herbicide Fact Sheet 2, 4-D: Toxicology, part 1," *Jour Pest Reform* 19, no. 1 (1999): 14–19.
62. US EPA, "List of Chemicals Evaluated for Carcinogenic Potential," Office of Pesticide Programs, 2000, accessed December 30, 2014, http://www.epa.gov/pesticides/health/cancerfs.htm#a.
63. US EPA, "2,4-D: Carcinogenicity Assessment for Lifetime Exposure," last updated October 31, 2014, http://epa.gov/ncea/iris/subst/0150.htm.
64. International Association for Research on Cancer (IARC), "IARC Classification of the Herbicide 2,4-D, Updated April 2010," May 4, 2010, accessed January 16, 2015, http://www.24d.org/backgrounders/body.aspx?pageID=36&contentID=129.

65. US EPA, "2,4-D RED Facts: Human Health Assessment – Toxicity," June 30, 2005, accessed December 30, 2014, http://www.epa.gov/oppsrrd1/reregistration/REDs/factsheets/24d_fs.htm.
66. Ibid.
67. Ismail Cakmak, Atilla Yazici, Yusuf Tutus, and Levent Ozturk, "Glyphosate Reduced Seed and Leaf Concentrations of Calcium, Manganese, Magnesium, and Iron in Non-Glyphosate Resistant Soybean," *European Journal of Agronomy* 31, no. 3 (2009): 114–119.
68. H. C. Steinrücken and N. Amrhein, "The Herbicide Glyphosate is a Potent Inhibitor of 5-Enolpyruvylshikimic Acid-3-Phosphate Synthase," *Biochemical and Biophysical Research Communications* 94, no. 4 (1980): 1207–1212.
69. US Patent 7,771,736, issued August 10, 2010.
70. Dwayne C. Savage, "Microbial Ecology of the Gastrointestinal Tract," *Annual Reviews in Microbiology* 31, no. 1 (1977): 107–133.
71. Sarkis K. Mazmanian, Cui Hua Liu, Arthur O. Tzianabos, and Dennis L. Kasper, "An Immunomodulatory Molecule of Symbiotic Bacteria Directs Maturation of the Host Immune System," *Cell* 122, no. 1 (2005): 107–118.
72. Awad A. Shehata, Wieland Schrödl, Alaa A. Aldin, Hafez M. Hafez, and Monika Krüger, "The Effect of Glyphosate on Potential Pathogens and Beneficial Members of Poultry Microbiota in Vitro," *Current Microbiology* 66, no. 4 (2013): 350–358.
73. Monika Krüger, Awad Ali Shehata, Wieland Schrödl, and Arne Rodloff. "Glyphosate Suppresses the Antagonistic Effect of *Enterococcus* spp. on *Clostridium botulinum*," *Anaerobe* 20 (2013): 74–78.
74. Caroline Cox, "Imazapyr," *Journal of Pesticide Reform* 16, no. 3 (Fall 1996).
75. Ibid.
76. Ibid.
77. A. Rahman, T. K. James, and P. Sanders, "Leaching and Movement of Imazapyr in Different Soil Types," in *Proceedings of the New Zealand Plant Protection Conference* (New Zealand Plant Protection Society Inc., 1993), 115–119.
78. CropLife International, "Biodiversity and the Plant Science Industry," July 2007, http://www.croplifeafrica.org/uploads/File/publications/260_PUB-BR_2007_07_13_Biodiversity_Case_Studies_-_Managing_natural_resources_sustainably.pdf.
79. Aaron G. Hagar and Dawn Nordby, "Herbicide Persistence and How to Test for Residues in Soils," in *2007 Illinois Agricultural Pest Management Handbook* (University of Illinois, 2007), http://ipm.illinois.edu/pubs/iapmh/15chapter.pdf.
80. Valerie Wilkinson and R. L. Lucas, "Effects of Herbicides on the Growth of Soil Fungi," *New Phytologist* 68, no. 3 (1969): 709–719.
81. Robert J. Kremer and Nathan E. Means, "Glyphosate and Glyphosate-Resistant Crop Interactions with Rhizosphere Microorganisms," *European Journal of Agronomy* 31, no. 3 (2009): 153–161.
82. T. R. Roberts, "Metabolic Pathways of Agrochemicals: Part 1: Herbicides and Plant Growth Regulators," (Cambridge, UK: The Royal Society of Chemistry, 1998), 396–399.
83. Food and Agriculture Organization, "Pesticide Residues in Food 2005: Evaluations Part 1 – Residues," joint meeting of the FAO Panel of Experts on Pesticide Residues in Food and the Environment and the WHO Core Assessment Group, Geneva, Switzerland, September 20-29, 2005, http://www.fao.org/fileadmin/templates/agphome/documents/Pests_Pesticides/JMPR/Evaluation05/2005Evaluation.pdf.
84. Robert Schwarcz, William O. Whetsell, and Richard M. Mangano, "Quinolinic Acid: an Endogenous Metabolite that Produces Axon-Sparing Lesions in Rat Brain," *Science* 219, no. 4582 (1983): 316–318.
85. University of California Agriculture and Natural Resources, "Herbicide Application Schools: Teaching Proper Sprayer Use and Calibration to Wildland Weed Warriors," accessed February 5, 2014, http://ucanr.edu/sites/statewideconference2013/files/165818.pdf.

86. Lisa-Natalie Anjozian, "The Tao of Treating Weeds: Reaching for Restoration in the Northern Rocky Mountains," *Fire Science Brief* 18 (2008):1–6.
87. Xia Ge, D. André d'Avignon, Joseph J. H. Ackerman, and R. Douglas Sammons, "Rapid Vacuolar Sequestration: the Horseweed Glyphosate Resistance Mechanism," *Pest Management Science* 66, no. 4 (2010): 345–348.
88. US Fish and Wildlife Service, "Managing Invasive Plants: Concepts, Principles, and Practices," last updated February 18, 2009, accessed January 21, 2014, http://www.fws.gov/invasives/stafftrainingmodule/methods/biological/introduction.html.
89. P. H. Dunn and L. A. Andres, "Entomopathogens Associated with Insects Used for Biological Control of Weeds," *Proceedings of the Fifth International Symposium on the Biological Control of Weeds* (1980): 241–246.
90. Dean E. Pearson and Ragan M. Callaway, "Indirect Effects of Host-Specific Biological Control Agents," *Trends in Ecology & Evolution* 18, no. 9 (2003): 456–461.
91. C. L. Campbell and J. P. McCaffrey, "Population Trends, Seasonal Phenology, and Impact of *Chrysolina Quadrigemina*, *C. hyperici* (Coleoptera: Chrysomelidae), and *Agrilus hyperici* (Coleoptera: Buprestidae) associated with *Hypericum perforatum* in Northern Idaho," *Environmental Entomology* 20, no. 1 (1991): 303–315.
92. Linda M. Wilson and Joseph P. McCaffrey, "Assessment of Biological Control of Exotic Broadleaf Weeds in Intermountain Rangelands," presented at the symposium on Ecology, Management, and Restoration of Intermountain Annual Rangelands, Boise, ID, May 18–22, 1992, from North Dakota State University Institutional Repository, accessed March 12, 2014, http://library.ndsu.edu/tools/dspace/load/?file=/repository/bitstream/handle/10365/2823/465mcc94.pdf?sequence=1.
93. Earl D. McCoy and J. H. Frank, "How Should the Risk Associated with the Introduction of Biological Control Agents be Estimated?" *Agricultural and Forest Entomology* 12, no. 1 (2010): 1–8.

Chapter 2: Getting to the Root of Invasive Species

1. Kathleen Sayce, *Introduced Cordgrass,* Spartina alterniflora *Loisel., in Salt Marshes and Tidelands of Willapa Bay, Washington,* US Fish and Wildlife Service, Willapa National Wildlife Refuge, Ilwaco, Washington,1988.
2. Paul Hedge, Lorne K. Kriwoken, and Kim Patten, "A review of Spartina management in Washington State, US," *Journal of Aquatic Plant Management* 41 (2003): 82–90.
3. Edwin D. Grosholz, Lisa A. Levin, Anna C. Tyler, and Carlos Neira, "Changes in Community Structure and Ecosystem Function Following *Spartina alterniflora* Invasion of Pacific Estuaries," in *Human Impacts on Salt Marshes: A Global Perspective* (Berkeley: University of California Press, 2009), 23–40.
4. Harry W. Wells, "The Fauna of Oyster Beds, with Special Reference to the Salinity Factor," *Ecological Monographs* (1961): 239–266.
5. "A State of Emergency" accessed February 22, 2015, http://brownmarsh.com/bms/1/index.htm.
6. Irving A. Mendelssohn and Nathan L. Kuhn, "Sediment Subsidy: Effects on Soil–Plant Responses in a Rapidly Submerging Coastal Salt Marsh," *Ecological Engineering* 21, no. 2 (2003): 115–128.
7. Robert I. Colautti and Hugh J. MacIsaac, "A Neutral Terminology to Define 'Invasive' Species," *Diversity and Distributions* 10, no. 2 (2004): 135–141.
8. US Department of Agriculture (USDA), "What is an Invasive Species?" accessed June 12, 2014, http://www.invasivespeciesinfo.gov/whatis.shtml.
9. US Congress Office of Technology Assessment, *Harmful Non-Indigenous Species in the United States,* US Government Printing Office, 1993.
10. Edwin D. Grosholz, Lisa A. Levin, Anna C. Tyler, and Carlos Neira, "Changes in Community Structure and Ecosystem Function Following *Spartina alterniflora* Invasion of Pacific Estuaries," in *Human Impacts on Salt Marshes: A Global Perspective* (Berkeley: University of California Press,

2009): 23–40; and Lisa A. Levin, Carlos Neira, and Edwin D. Grosholz, "Invasive Cordgrass Modifies Wetland Trophic Function," *Ecology* 87, no. 2 (2006): 419–432.

11. Edwin D. Grosholz, Lisa A. Levin, Anna C. Tyler, and Carlos Neira, "Changes in Community Structure and Ecosystem Function Following *Spartina alterniflora* Invasion of Pacific Estuaries," in *Human Impacts on Salt Marshes: A Global Perspective* (Berkeley: University of California Press, 2009): 23–40.
12. Lisa A. Levin, Carlos Neira, and Edwin D. Grosholz, "Invasive Cordgrass Modifies Wetland Trophic Function," *Ecology* 87, no. 2 (2006): 419–432.
13. For example, Leonard A. Brennan, Michael A. Finger, Joseph B. Buchanan, Charles T. Schick, and Steven G. Herman, "Stomach Contents of Dunlins Collected in Western Washington," *Northwestern Naturalist* (1990): 99–102; Susan K. Skagen and Fritz L. Knopf, "Migrating Shorebirds and Habitat Dynamics at a Prairie Wetland Complex," *The Wilson Bulletin* (1994): 91–105; and Craig A. Davis and Loren M. Smith, "Foraging Strategies and Niche Dynamics of Coexisting Shorebirds at Stopover Sites in the Southern Great Plains," *The Auk* 118, no. 2 (2001): 484–495.
14. Joseph B. Buchanan, "Spartina Invasion of Pacific Coast Estuaries in the United States: Implications for Shorebird Conservation," *Bulletin-Wader Study Group* 100 (2003): 47–49.
15. Mark A. Davis, "Researching Invasive Species 50 Years after Elton: A Cautionary Tale," in *Fifty Years of Invasion Ecology: The Legacy of Charles Elton*, ed. David M. Richardson (Hoboken, NJ: John Wiley & Sons, 2011), 274.
16. Kim Patten and Carol O'Casey, "Use of Willapa Bay, Washington, by Shorebirds and Waterfowl After Spartina Control Efforts," *Journal of Field Ornithology* 78, no. 4 (2007): 395–400.
17. Ibid.
18. Jared R. Parks, "Shorebird Use of Smooth Cordgrass (*Spartina alterniflora*) Meadows in Willapa Bay, Washington," master's thesis, The Evergreen State College, 2006.
19. Cory Tyler Overton, "Tidally-Induced Limits to California Clapper Rail Ecology in San Francisco Bay Salt Marshes," PhD dissertation, University of California Davis, 2014.
20. Jen McBroom, "California Clapper Rail Surveys for the San Francisco Estuary Invasive Spartina Project," Olofson Environmental, Inc., November 2013.
21. David A. Armstrong and Raymond E. Millemann, "Effects of the Insecticide Carbaryl on Clams and Some Other Intertidal Mud Flat Animals," *Journal of the Fisheries Board of Canada* 31, no. 4 (1974): 466–470.
22. US EPA, "Environmental Fate and Ecological Risk Assessment for the Re-Registration of Carbaryl," March 2003, http://www.epa.gov/espp/litstatus/effects/carb-riskass.pdf, 26.
23. Caroline Cox, "Insecticide Fact Sheet: Imidacloprid," *Journal of Pesticide Reform* 21, no. 1 (2001): 15–21.
24. Canadian Council of Ministers of the Environment, "Canadian Water Quality Guidelines: Imidacloprid Scientific Supporting Document," 2007, http://www.ccme.ca/files/Resources/supporting_scientific_documents/imidacloprid_ssd_1388.pdf.
25. Kim Patten, "Evaluating Imazapyr in Aquatic Environments: Searching for Ways to Stem the Tide of Aquatic Weeds," *Agrichemical and Environmental News* 205, May 2003.
26. Kim Patten, "Spartina Control in the Willapa Bay – A Success Story," 2006, http://longbeach.wsu.edu/spartina/documents/spartinacontrolinwillapabayasuccessstory-talkatapms2006.pdf.
27. Washington State Department of Agriculture (WSDA), "Spartina Eradication Program: 2012 Progress Report," March 2013, http://agr.wa.gov/PlantsInsects/Weeds/Spartina/docs/SpartinaReport2012.pdf.
28. Washington Invasive Species Council, "2010: Report to the Legislature on the Future of the Council," 2010, http://www.invasivespecies.wa.gov/documents/2010Annual_Report.pdf.
29. WSDA, "2012 Progress Report."
30. Daniel Simberloff, "The Politics of Assessing Risk for Biological Invasions: the USA as a Case Study," *Trends in Ecology & Evolution* 20, no. 5 (2005): 216–222.

31. Mandy Tu, Callie Hurd, and John M. Randall, "Weed Control Methods Handbook: Tools & Techniques for use in Natural Areas," 2001, http://www.invasive.org/gist/products/handbook/07.herbicideguidelines.pdf, 5.1.
32. Matthew J. Rinella, Bruce D. Maxwell, Peter K. Fay, Theodore Weaver, and Roger L. Sheley, "Control Effort Exacerbates Invasive-Species Problem," *Ecological Applications* 19, no. 1 (2009): 155–162.
33. Kim Patten and Carol O'Casey, "Shorebird, Waterfowl and Birds of Prey Usage of Willapa Bay in Response to *Spartina* Control Efforts," progress report to the USFW – Willapa Wildlife Refuge, 2006, http://longbeach.wsu.edu/spartina/documents/shorebirdmonitoringprogressreport porterptwillapanationalwildliferefugejanuary2005.pdf, 24.
34. Coalition to Protect Puget Sound Habitat, "Re: Appeal of the Washington State Noxious Weed Control Board Denial of Rule-Making to Amend Petition Filed by the Coalition to Protect Puget Sound Habitat and Robert Kavanaugh," January 27, 2014, http://coalitiontoprotectpuget soundhabitat.org/wp-content/uploads/2014/01/eelgrass-petition.pdf.
35. Grosholz et al., "Changes in Community Structure."
36. Ludwig Von Bertalanffy, *General System Theory: Foundation, Development, Applications* (New York: George Braziller, 1968), 31.
37. Thomas S. Kuhn, *The Structure of Scientific Revolutions* (Chicago: University of Chicago Press, 1970), 150.
38. Thomas E. Dahl and Gregory J. Allord, "History of Wetlands in the Conterminous United States," USGS Water-Supply Paper 2425, 1996, 19–26.
39. H. Galbraith, R. Jones, R. Park, J. Clough, S. Herrod-Julius, B. Harrington, and G. Page, "Global Climate Change and Sea Level Rise: Potential Losses of Intertidal Habitat for Shorebirds," *Waterbirds* 25, no. 2 (2002): 173–183.
40. Steven E. Schwarzbach, Joy D. Albertson, and Carmen M. Thomas, "Effects of Predation, Flooding, and Contamination on Reproductive Success of California Clapper Rails (*Rallus longirostris obsoletus*) in San Francisco Bay," *The Auk* 123, no. 1 (2006): 45–60.
41. Paul Baumann, "The Shrinkage of a Water Surface: San Francisco Bay," 2001, accessed October 12, 2014, http://www.oneonta.edu/faculty/baumanpr/geosat2/Shrinking_Bay/Shrinking_Bay.htm.
42. Guy Gelfenbaum, Christopher R. Sherwood, Curt D. Peterson, George M. Kaminsky, Maarten Buijsman, David C. Twichell, Peter Ruggiero, Ann E. Gibbs, and Christopher Reed, "The Columbia River Littoral Cell: a Sediment Budget Overview," *Proceedings of Coastal Sediments*, 99 (1999): 1660–1675.
43. Daniel E. Campbell, Hong-Fang Lu, George A. Know, and Howard T. Odum, "Maximizing Empower on a Human-Dominated Planet: The Role of Exotic *Spartina*," *Ecological Engineering* 35 (2009): 463–486.
44. Leopold, Aldo, *Round River: From the Journals of Aldo Leopold* (Oxford: Oxford University Press, 1993), 165.

Chapter 3: Thinking Like an Ecosystem

1. Michael J. Newman and Robert T. Rood, "Implications of Solar Evolution for the Earth's Early Atmosphere." *Science* 198, no. 4321 (1977): 1035–1037.
2. Lynn Margulis, *Symbiotic Planet: A New Look at Evolution* (New York: Basic Books, 2008), 119.
3. Marcel Rejmánek, David M. Richardson, and P. Pyšek, "Plant Invasions and Invasibility of Plant Communities," *Vegetation Ecology* (2005): 332–355.
4. Mark A. Davis and Melissa Pelsor, "Experimental Support for a Resource-Based Mechanistic Model of Invasibility," *Ecology Letters* 4, no. 5 (2001): 421–428.
5. Donella H. Meadows, *Thinking in Systems: A Primer* (White River Junction, VT: Chelsea Green Publishing, 2008), 16.

6. US Bureau of Reclamation, "Colorado River System: Consumptive Uses and Losses Report," Revised September 2012, http://www.usbr.gov/uc/library/envdocs/reports/crs/pdfs/cul2001-05.pdf, iv.
7. Robert W. Adler, *Restoring Colorado River Ecosystems: A Troubled Sense of Immensity* (Washington, DC: Island Press, 2007), 215
8. Colorado River Basin Salinity Control Forum, "2011 Review: Water Quality Standards for Salinity Colorado River System," October 2011, http://coloradoriversalinity.org/docs/2011%20REVIEW-October.pdf.
9. Edmund D. Andrews, "Sediment Transport in the Colorado River Basin," in *Colorado River Ecology and Dam Management: Proceedings of a Symposium May 24–25, 1990 Santa Fe, New Mexico* (Washington, DC: National Academies Press, 1991): 54.
10. John C. Dohrenwend, "Rapid Progradation of the Colorado and San Juan Deltas into Lake Powell Reservoir, July 2002 to January 2004," in *Geological Society of America Abstracts with Programs* 36, no. 4 (2004): 15.
11. Living Rivers, Colorado Riverkeeper, "The One Dam Solution: Preliminary Report to the Bureau of Reclamation on Proposed Reoperation Strategies for Glen Canyon and Hoover Dam under Low Water Conditions," 2005, http://www.livingrivers.org/pdfs/TheOne-DamSolution.pdf.
12. Richard Ingebretsen, PhD, Glen Canyon Institute, e-mail message to author, February 26, 2015.
13. Hugo G. Hidalgo, Thomas C. Piechota, and John A. Dracup, "Alternative Principal Components Regression Procedures for Dendrohydrologic Reconstructions," *Water Resources Research* 36, no. 11 (2000): 3241–3249.
14. Tim P. Barnett and David W. Pierce, "When Will Lake Mead Go Dry?" *Water Resources Research* 44, no. 3 (2008).
15. National Park Service, "Grand Canyon's Native Fish," accessed January 1, 2015, http://www.nps.gov/grca/naturescience/fish-native.htm.
16. Julie Stromberg, "Dynamics of Fremont Cottonwood (*Populus fremontii*) and Saltcedar (*Tamarix chinensis*) Populations along the San Pedro River, Arizona," *Journal of Arid Environments* 40, no. 2 (1998): 133–155.
17. Mark K. Sogge, Susan J. Sferra, and Eben H. Paxton, "Tamarix as Habitat for Birds: Implications for Riparian Restoration in the Southwestern United States," *Restoration Ecology* 16, no. 1 (2008): 146–154.
18. Becky K. Kerns, Bridgett J. Naylor, Michelle Buonopane, Catherine G. Parks, and Brendan Rogers, "Modeling Tamarisk (*Tamarix* spp.) Habitat and Climate Change Effects in the Northwestern United States," *Invasive Plant Science and Management* 2, no. 3 (2009): 200–215.
19. "Salt Cedar and Russian Olive Control Demonstration Act," Public Law 109-320, October 11, 2006, http://www.gpo.gov/fdsys/pkg/PLAW-109publ320/pdf/PLAW-109publ320.pdf.
20. US Department of Agriculture (USDA) Forest Service, "Field Guide for Managing Salt Cedar," 2010, http://www.fs.usda.gov/Internet/FSE_DOCUMENTS/stelprdb5180537.pdf.
21. Joseph M. Di Tomaso, "Impact, Biology, and Ecology of Saltcedar (*Tamarix* spp.) in the Southwestern United States," *Weed Technology* 12 (1998): 326–336.
22. Pamela L. Nagler, Russell L. Scott, Craig Westenburg, James R. Cleverly, Edward P. Glenn, and Alfredo R. Huete, "Evapotranspiration on Western US Rivers Estimated using the Enhanced Vegetation Index from MODIS and Data from Eddy Covariance and Bowen Ratio Flux Towers," *Remote Sensing of Environment* 97, no. 3 (2005): 337–351.
23. Charles R. Hart, Larry D. White, Alyson McDonald, and Zhuping Sheng, "Saltcedar Control and Water Salvage on the Pecos River, Texas, 1999–2003," *Journal of Environmental Management* 75, no. 4 (2005): 399–409.
24. Brent C. Christner, "Cloudy with a Chance of Microbes," *Microbe Magazine* 7, no. 2 (Feb. 2012): 70–75.
25. Christine Dell'Amore, "Rainmaking Bacteria Ride Clouds to 'Colonize' Earth," *National Geographic News,* January 12, 2009.

26. Noga Qvit-Raz, Edouard Jurkevitch, and Shimshon Belkin, "Drop-Size Soda Lakes: Transient Microbial Habitats on a Salt-Secreting Desert Tree," *Genetics* 178, no. 3 (2008): 1615–1622.
27. Joseph H. Connell, "Diversity in Tropical Rain Forests and Coral Reefs: High Diversity of Trees and Corals is Maintained only in a Nonequilibrium State," *Science* 199, no. 4335 (1978): 1302–1310.
28. Joydeep Bhattacharjee, John P. Taylor, and Loren M. Smith, "Controlled Flooding and Staged Drawdown for Restoration of Native Cottonwoods in the Middle Rio Grande Valley, New Mexico, USA," *Wetlands* 26, no. 3 (2006): 691–702.
29. Colin R. Townsend, Mike R. Scarsbrook, and Sylvain Dolédec, "The Intermediate Disturbance Hypothesis, Refugia, and Biodiversity in Streams," *Limnology and Oceanography* 42, no. 5 (1997): 938–949.
30. David P. Tickner, Penelope G. Angold, Angela M. Gurnell, and J. Owen Mountford, "Riparian Plant Invasions: Hydrogeomorphological Control and Ecological Impacts," *Progress in Physical Geography* 25, no. 1 (2001): 22–52.
31. Gregory T. Munger, "Lonicera japonica," in *Fire Effects Information System,* [Online]. US Department of Agriculture, Forest Service, Rocky Mountain Research Station, Fire Sciences Laboratory (Producer), accessed April 12, 2014, http://www.fs.fed.us/database/feis/plants/vine/lonjap/all.html; John S. Kush, Ralph S. Meldahl, and William D. Boyer, "Understory Plant Community Response to Season of Burn in Natural Longleaf Pine Forests," in *Fire and Forest Ecology: Innovative Silviculture and Vegetation Management. Tall Timbers Fire Ecology Conference Proceedings*, ed. W. K. Moser (Tallahassee, FL: Tall Timbers Research Station, 2000), 32–39.
32. F. A. Bazzaz, "The Physiological Ecology of Plant Succession," *Annual Review of Ecology and Systematics* 10, no. 1 (1979): 351–371.
33. Henry A. Gleason, "The Individualistic Concept of the Plant Association," *Bulletin of the Torrey Botanical Club* 53, no. 1 (1926): 26.
34. Robert M. May, "Thresholds and Breakpoints in Ecosystems with a Multiplicity of Stable States," *Nature* 269, no. 5628 (1977): 471–477.
35. Jane A. Catford, Curtis C. Daehler, Helen T. Murphy, Andy W. Sheppard, Britta D. Hardesty, David A. Westcott, Marcel Rejmánek, Peter J. Bellingham, Jan Pergl, Carol C. Horvitz, and Philip E. Hulme, "The Intermediate Disturbance Hypothesis and Plant Invasions: Implications for Species Richness and Management," *Perspectives in Plant Ecology, Evolution and Systematics* 14, no. 3 (2012): 231–241.
36. Habacuc Flores-Moreno, Fiona J. Thomson, David I. Warton, and Angela T. Moles, "Are Introduced Species Better Dispersers Than Native Species? A Global Comparative Study of Seed Dispersal Distance," *PloS one* 8, no. 6 (2013): e68541.
37. Including studies like Jeffrey D. Corbin and Carla M. D'Antonio, "Gone but Not Forgotten? Invasive Plants' Legacies on Community and Ecosystem Properties," *Invasive Plant Science and Management* 5, no. 1 (2012): 117–124; Peter M. Vitousek and Lawrence R. Walker, "Biological Invasion by Myrica faya in Hawai'i: Plant Demography, Nitrogen Fixation, Ecosystem Effects," *Ecological Monographs* 59, no. 3 (1989): 247–265; and Joan G. Ehrenfeld, "Ecosystem Consequences of Biological Invasions," *Annual Review of Ecology, Evolution, and Systematics* 41 (2010): 59–80.
38. Crawford S. Holling, "The Resilience of Terrestrial Ecosystems: Local Surprise and Global Change," in *Sustainable Development of the Biosphere*, ed. W. C. Clark and R. E. Munn (Cambridge, UK: Cambridge University Press, 1986), 292–317.
39. NatureServe, "Invasive Species," accessed November 9, 2014, http://www.natureserve.org/biodiversity-science/conservation-topics/invasive-species.
40. USDA Forest Service, "Invasive Species Research Model," accessed November 6, 2014, http://www.fs.fed.us/rm/wildlife-terrestrial/invasive-species/projects/research-model.php.
41. Bill Mollison, *Permaculture: A Designer's Manual* (Tasmania: Tagari Press, 1988), 33.

Chapter 4: A Matter of Time

1. Stephen Hasiotis, Sandra Brake, and Kathy Dannelly, "An Ecosystem Engineer in Acid Mine Drainage Systems, Western Indiana: Euglena Mutabilis, an Oxygenic, Photosythesizing, Stromatolite-Building Protozoan Making Life Possible for Others," *Geomicrobiology Posters*. No. 24 (2001).
2. Michael Bau and Peter Möller, "Rare Earth Element Systematics of the Chemically Precipitated Component in Early Precambrian Iron Formations and the Evolution of the Terrestrial Atmosphere-hydrosphere-lithosphere System," *Geochimica et Cosmochimica Acta* 57, no. 10 (1993): 2239–2249.
3. Lynn Margulis. *Microcosmos: four billion years of evolution from our microbial ancestors*. (Berkeley: University of California Press, 1986), 29.
4. Anita Narwani, Markos A. Alexandrou, Todd H. Oakley, Ian T. Carroll, and Bradley J. Cardinale, "Experimental Evidence that Evolutionary Relatedness Does Not Affect the Ecological Mechanisms of Coexistence in Freshwater Green Algae," *Ecology Letters* 16, no. 11 (2013): 1373–1381.
5. Frantisek Baluska and Stefano Mancuso, "Microorganism and Filamentous Fungi Drive Evolution of Plant Synapses," *Frontiers in Cellular and Infection Microbiology* 3 (2013): 44.
6. Akie Sato, Herbert Tichy, Colm O'Huigin, Peter R. Grant, B. Rosemary Grant, and Jan Klein, "On the Origin of Darwin's Finches," *Molecular Biology and Evolution* 18, no. 3 (2001): 299–311.
7. Charles Darwin, *The Origin of Species by Means of Natural Selection: or, the Preservation of Favored Races in the Struggle for Life* (New York: Penguin, 1968), 84.
8. Samantha J. Price-Rees, Gregory P. Brown, and Richard Shine, "Interacting Impacts of Invasive Plants and Invasive Toads on Native Lizards," *The American Naturalist* 179, no. 3 (2012): 413–422.
9. Courtney L. J. Rowe and Elizabeth A. Leger, "Competitive Seedlings and Inherited Traits: A Test of Rapid Evolution of *Elymus multisetus* (big squirreltail) in Response to Cheatgrass Invasion," *Evolutionary Applications* 4, no. 3 (2011): 485–498.
10. David S. Wilcove, David Rothstein, Jason Dubow, Ali Phillips, and Elizabeth Losos, "Quantifying Threats to Imperiled Species in the United States," *BioScience* (1998): 607–615.
11. James H. Miller, Erwin B. Chambliss, Christopher M. Oswalt, "Maps of Occupation and Estimates of Acres Covered by Nonnative Invasive Plants in Southern Forests," using SRS FIA data posted on March 15, 2008. Accessed March 27, 2014, http://www.invasive.org/fiamaps/.
12. United States Department of Agriculture (USDA), "Crop Production: 2014 Summary" (January 2015)," http://www.usda.gov/nass/PUBS/TODAYRPT/cropan15.pdf.
13. David M. Richardson, Nicky Allsopp, Carla M. D'Antonio, Suzanne J. Milton, and Marcel Rejmánek, "Plant Invasions – the Role of Mutualisms," *Biological Reviews* 75, no. 1 (2000): 65–93.
14. Jason M. Gleditsch and Tomás A. Carlo, "Fruit Quantity of Invasive Shrubs Predicts the Abundance of Common Native Avian Frugivores in Central Pennsylvania," *Diversity and Distributions* 17, no. 2 (2011): 244–253.
15. Kristin I. Powell, Jonathan M. Chase, and Tiffany M. Knight, "A Synthesis of Plant Invasion Effects on Biodiversity Across Spatial Scales," *American Journal of Botany* 98, no. 3 (2011): 539–548.
16. Lizette F. Cook and Catherine A. Toft, "Dynamics of Extinction: Population Decline in the Colonially Nesting Tricolored Blackbird *Agelaius tricolor*," *Bird Conservation International* 15, no. 01 (2005): 73–88.
17. Ibid.
18. David M. Raup and J. John Sepkoski Jr., "Mass Extinctions in the Marine Fossil Record," *Science* 215, no. 4539 (1982): 1501–1503.
19. Daniel J. Lehrmann Jahandar Ramezani, Samuel A. Bowring, Mark W. Martin, Paul Montgomery, Paul Enos, Jonathan L. Payne, Michael J. Orchard, Wang Hongmei, and Wei Jiayong, "Timing of Recovery from the End-Permian Extinction: Geochronologic and

Biostratigraphic Constraints from South China," *Geology* 34, no. 12 (2006): 1053–1056; and Sarda Sahney and Michael J. Benton, "Recovery from the Most Profound Mass Extinction of All Time," *Proceedings of the Royal Society B: Biological Sciences* 275, no. 1636 (2008): 759–765.

20. E. Chivian and A. Bernstein (eds.), *Sustaining Life: How Human Health Depends on Biodiversity* (New York: Oxford University Press, 2008).
21. Jessica Gurevitch and Dianna K. Padilla, "Are Invasive Species a Major Cause of Extinctions?" *Trends in Ecology & Evolution* 19, no. 9 (2004): 470–474.
22. John Kartesz, e-mail message to author, September 20, 2013.
23. Ibid.
24. Richard Ogutu-Ohwayo, "The Decline of the Native Fishes of Lakes Victoria and Kyoga (East Africa) and the Impact of Introduced Species, Especially the Nile Perch, *Lates niloticus*, and the Nile Tilapia, *Oreochromis niloticus*," *Environmental Biology of Fishes* 27, no. 2 (1990): 81–96.
25. Georges Raab, *1994 IUCN Red List of Threatened Animals*, (Gland, Switzerland: IUCN, 1994), xviii.
26. J. D. Fridley and D. F. Sax, "The Imbalance of Nature: Revisiting a Darwinian Framework for Invasion Biology," *Global Ecology and Biogeography* 23 (2014): 1157–1166.
27. Gordon H. Rodda, Thomas H. Fritts, and Paul J. Conry, "Origin and Population Growth of the Brown Tree Snake, *Boiga irregularis*, on Guam," in George P. L. Walker, George, Elizabeth Ann Powell, Kathleen A. Lohse, Dennis Nullet, Peter M. Vitousek, Timothy J. Motley, Stephen L. Coles et al., "A Quarterly Devoted to the Biological and Physical Sciences of the Pacific Region," *Abstracts of Papers. Sixteenth Annual Albert L. Tester Memorial* 90 (1991).
28. Smithsonian National Zoological Park, "Where Have All the Birds of Guam Gone?" accessed January 17, 2015, http://nationalzoo.si.edu/Animals/Birds/Facts/fact-guambirds.cfm.
29. Jenkins, J. Mark, "Natural History of the Guam Rail," *Condor* (1979): 404–408.
30. Michael E. Wheeler, "Man-Wildlife Interactions on Guam," *Transactions of the Western Section of the Wildlife Society* 16 (1980), http://www.tws-west.org/transactions/Wheeler.pdf.
31. Daniel Simberloff, "Extinction-Proneness of Island Species-Causes and Management Implications," *Raffles Bulletin of Zoology* 48, no. 1 (2000): 1–9.
32. Thomas H. Fritts, "Demise of an Insular Avifauna: the Brown Tree Snake on Guam," *Transactions of the Western Section of the Wildlife Society* 24 (1988): 31–37.
33. Ibid.
34. J. Mark Jenkins, "Natural History of the Guam Rail," *Condor* (1979): 404–408.
35. J. Mark Jenkins, *The Native Forest Birds of Guam* (Washington, DC: American Ornithologists' Union, 1983).
36. US Fish and Wildlife Service, "Scientific Evaluation of the Status of the Northern Spotted Owl," September 2004, https://www.fws.gov/oregonfwo/species/Data/NorthernSpottedOwl/BarredOwl/Documents/CourtneyEtAl2004.pdf: 7–4.
37. US Fish and Wildlife Service, "U.S. Fish and Wildlife Announces Final Decision for Barred Owl Removal Experiment," news release September 10, 2013, http://www.fws.gov/oregonfwo/ExternalAffairs/News/Documents/BarredOwlRODSep2013.pdf.
38. Craig Welch, "The Spotted Owl's New Nemesis," *Smithsonian*, January 2009.
39. Gurevitch and Padilla, "Are Invasive Species a Major Cause of Extinctions?" 470.
40. Chris D. Thomas, Alison Cameron, Rhys E. Green, Michel Bakkenes, Linda J. Beaumont, Yvonne C. Collingham, Barend F. N. Erasmus, et al., "Extinction Risk from Climate Change," *Nature* 427, no. 6970 (2004): 145–148.
41. Paul N. Pearson and Martin R. Palmer, "Atmospheric Carbon Dioxide Concentrations over the Past 60 Million Years," *Nature* 406, no. 6797 (2000): 695–699.
42. Steven C. Sherwood, Sandrine Bony, and Jean-Louis Dufresne, "Spread in Model Climate Sensitivity Traced to Atmospheric Convective Mixing," *Nature* 505, no. 7481 (2014): 37–42.

43. Dianne Dumanoski, *The End of the Long Summer: Why We Must Remake Our Civilization to Survive on a Volatile Earth* (New York: Random House LLC, 2009).
44. Gerald E. Rehfeldt, Nicholas L. Crookston, Marcus V. Warwell, and Jeffrey S. Evans, "Empirical Analyses of Plant–Climate Relationships for the Western United States," *International Journal of Plant Sciences* 167, no. 6 (2006): 1123–1150.
45. Liette Vasseur and Catherine Potvin, "Natural Pasture Community Response to Enriched Carbon Dioxide Atmosphere," *Plant Ecology* 135, no. 1 (1998): 31–41.
46. Susan C. L. McDowell, "Photosynthetic Characteristics of Invasive and Noninvasive Species of *Rubus* (Rosaceae)," *American Journal of Botany* 89 (September 2002): 1431–1438
47. Liying Song Jinrong Wu, Changhan Li, Furong Li, Shaolin Peng, and Baoming Chen, "Different Responses of Invasive and Native Species to Elevated CO_2 Concentration," *Acta oecologica* 35, no. 1 (2009): 128–135.
48. Paul N. Pearson and Martin R. Palmer, "Atmospheric Carbon Dioxide Concentrations over the Past 60 Million Years," *Nature* 406, no. 6797 (2000): 695–699.
49. E. N. Speelman, M. M. L. Van Kempen, J. Barke, H. Brinkhuis, G. J. Reichart, A. J. P. Smolders, J. G. M. Roelofs, et al., "The Eocene Arctic Azolla Bloom: Environmental Conditions, Productivity and Carbon Drawdown," *Geobiology* 7, no. 2 (2009): 155–170.
50. Miguel A. Altieri and Clara I. Nicholls, "Soil Fertility Management and Insect Pests: Harmonizing Soil and Plant Health in Agroecosystems," *Soil and Tillage Research* 72, no. 2 (2003): 203–211.
51. Ignacio Quintero and John J. Wiens, "Rates of Projected Climate Change Dramatically Exceed Past Rates of Climatic Niche Evolution among Vertebrate Species," *Ecology Letters* 16, no. 8 (2013): 1095–1103.
52. B. Huntley and H. J. B. Birks, An Atlas of Past and Present Pollen Maps for Europe: 0–13,000 Years Ago (Cambridge, UK: Cambridge University Press, 1983).
53. Benjamin G. Freeman and Alexandra M. Class Freeman, "Rapid Upslope Shifts in New Guinean Birds Illustrate Strong Distributional Responses of Tropical Montane Species to Global Warming," *Proceedings of the National Academy of Sciences* 111, no. 12 (2014): 4490–4494.
54. Jillian M. Mueller and Jessica J. Hellmann, "An Assessment of Invasion Risk from Assisted Migration," *Conservation Biology* 22, no. 3 (2008): 562–567.
55. Mark Vellend, Luke J. Harmon, Julie L. Lockwood, Margaret M. Mayfield, A. Randall Hughes, John P. Wares, and Dov F. Sax, "Effects of Exotic Species on Evolutionary Diversification," *Trends in Ecology & Evolution* 22, no. 9 (2007): 481–488.
56 Dov F. Sax, Steven D. Gaines, and James H. Brown, "Species Invasions Exceed Extinctions on Islands Worldwide: A Comparative Study of Plants and Birds," *The American Naturalist* 160, no. 6 (2002): 766–783.
57. Paul C. Davis, "Conifer Trees Thrive in New Zealand," *California Agriculture*, (1980): 4–6.
58. Chris D. Thomas, "Translocation of Species, Climate Change, and the End of Trying to Recreate Past Ecological Communities," *Trends in Ecology & Evolution* 26, no. 5 (2011): 216–221.
59. Xinbin Duan, Shaoping Liu, Mugui Huang, Shunling Qiu, Zhihua Li, Ke Wang, and Daqing Chen, "Changes in Abundance of Larvae of the Four Domestic Chinese Carps in the Middle Reach of the Yangtze River, China, Before and After Closing of the Three Gorges Dam," in *Chinese Fishes* (Netherlands: Springer, 2009), 13–22.
60. Fan Li, "Fifth Century BC The Chinese Fish Culture Classic," translated by T.S.Y. Koo, Reprinted as "Contribution 489, Chesapeake Biological Laboratory, University of Maryland, Solomons, Maryland," in Shao Wen Ling, *Aquaculture in Southeast Asia* (Seattle: University of Washington Press, 1977), 103.
61. Federal Register, "Injurious Wildlife Species; Silver Carp and Largescale Silver Carp," *Federal Register* 72, no. 131 (2007).

62. Brian S. Ickes, "Asian Carp: Perspectives from Both Sides of the Planet," May 18, 2012, http://www.nps.gov/miss/naturescience/upload/IckesPresent-051812.pdf.
63. Julie Turgeon and Louis Bernatchez, "Reticulate Evolution and Phenotypic Diversity in North American Ciscoes, *Coregonus* spp. (Teleostei: Salmonidae): Implications for the Conservation of an Evolutionary Legacy," *Conservation Genetics* 4, no. 1 (2003): 67–81.
64. David Bickford, David J. Lohman, Navjot S. Sodhi, Peter K. L. Ng, Rudolf Meier, Kevin Winker, Krista K. Ingram, and Indraneil Das, "Cryptic Species as a Window on Diversity and Conservation," *Trends in Ecology & Evolution* 22, no. 3 (2007): 148–155.
65. Joel Hood, "State's Fish Kill in Sanitary and Ship Canal Yields Only 1 Asian Carp," *Chicago Tribune*, December 4, 2009.
66. Sophia Tareen, "Illinois Launches Asian Carp Anti-Hunger Program," Associated Press, September 23, 2011.
67. Bill Mollison, *Permaculture: A Designer's Manual* (Tasmania: Tagari Press, 1988).

Chapter 5: Problems into Solutions

1. Millennium Ecosystem Assessment, "Ecosystems and Human Well-Being," World Resources Institute, 2005, http://www.millenniumassessment.org/documents/document.356.aspx.pdf.
2. Heather Charles and Jeffrey S. Dukes, "13 Impacts of Invasive Species on Ecosystem Services," *Ecological Studies* 193 (2007): 217.
3. Peter M. Vitousek and Lawrence R. Walker, "Biological Invasion by *Myrica faya* in Hawai'i: Plant Demography, Nitrogen Fixation, Ecosystem Effects," *Ecological Monographs* 59, no. 3 (1989): 247–265.
4. G. H. Aplet, "Alteration of Earthworm Community Biomass by the Alien *Myrica faya* in Hawai'i," *Oecologia* 82, no. 3 (1990): 414–416.
5. Dana Cordell, Jan-Olof Drangert, and Stuart White, "The Story of Phosphorus: Global Food Security and Food for Thought," *Global Environmental Change* 19, no. 2 (2009): 292–305.
6. B. Wang and Y-L. Qiu, "Phylogenetic Distribution and Evolution of Mycorrhizas in Land Plants," *Mycorrhiza* 16, no. 5 (2006): 299–363.
7. Anne Pringle, James D. Bever, Monique Gardes, Jeri L. Parrent, Matthias C. Rillig, and John N. Klironomos, "Mycorrhizal Symbioses and Plant Invasions," *Annual Review of Ecology, Evolution, and Systematics* 40 (2009): 699–715.
8. Andrea S. Thorpe, Vince Archer, and Thomas H. DeLuca, "The Invasive Forb, *Centaurea maculosa*, Increases Phosphorus Availability in Montana Grasslands," *Applied Soil Ecology* 32, no. 1 (2006): 118–122.
9. Richy J. Harrod and Ronald J. Taylor, "Reproduction and Pollination Biology of Centaurea and Acroptilon Species, with Emphasis on *C. diffusa*," *Northwest Science* 69, no. 2 (1995): 97–105.
10. Valerie A. Miller, "Knapweed as Forage for Big Game in the Kootenays," in *Range Weeds Revisited: Proceedings of a Symposium: A 1989 Pacific Northwest Range Management Short Course*, ed. Ben F. Roche, Cindy Talbott Roche January 24–26, Spokane, WA, Washington State University, Department of Natural Resource Sciences, Cooperative Extension, 1989, 35–37.
11. Bret E. Olson, Roseann T. Wallander, and John R. Lacey, "Effects of Sheep Grazing on a Spotted Knapweed-Infested Idaho Fescue Community," *Journal of Range Management* (1997): 386–390.
12. Arthur M. Shapiro, "The Californian Urban Butterfly Fauna Is Dependent on Alien Plants," *Diversity and Distributions* 8, no. 1 (2002): 31–40.
13. Anthony Ricciardi and Joseph B. Rasmussen, "Extinction Rates of North American Freshwater Fauna," *Conservation Biology* 13, no. 5 (1999): 1220–1222.
14. Paul L. Klerks, Peter C. Fraleigh, and Robert Sinsabaugh, "The Impact of Zebra Mussels on the Dynamics of Heavy Metals," *Technical Summary OHSU-TB-055, Ohio Sea Grant College Program* (2001).

15. Maria Alexandra Bighiu, "Zebra Mussels (*Dreissena polymorpha*) for Accessing Microbial Contamination and Antibiotic Resistant Bacteria in Freshwaters," (Thesis, Swedish University of Agricultural Sciences, 2012).
16. Scott A. Petrie and Richard W. Knapton, "Rapid increase and subsequent decline of zebra and quagga mussels in Long Point Bay, Lake Erie: possible influence of waterfowl predation," *Journal of Great Lakes Research* 25, no. 4 (1999): 772-782.
17. Hugh J. MacIsaac, "Potential abiotic and biotic impacts of zebra mussels on the inland waters of North America," *American Zoologist* 36, no. 3 (1996): 287-299.
18. Carrie E. H. Kissman, Lesley B. Knoll, and Orlando Sarnelle, "Dreissenid Mussels (*Dreissena polymorpha* and *Dreissena bugensis*) reduce Microzooplankton and Macrozooplankton Biomass in Thermally Stratified Lakes," *Limnology and Oceanography* 55, no. 5 (2010): 1851–1859.
19. Douglas Endicott, Russell G. Kreis Jr., Lauri Mackelburg, and Dean Kandt, "Modeling PCB Bioaccumulation by the Zebra Mussel (*Dreissena polymorpha*) in Saginaw Bay, Lake Huron," *Journal of Great Lakes Research* 24, no. 2 (1998): 411–426.
20. US Environmental Protection Agency (US EPA), "The Effect of Zebra Mussels on Cycling and Potential Bioavailability of PCBs: Case Study of Saginaw Bay: Section 6 – Conclusion and Recommendations," accessed May 16, 2014, http://www.epa.gov/greatlakes/invasive/zmussels/sec6.html.
21. V. Ramesh, C. Karunakaran, and A. Rajendran, "Optimization of Submerged Culture Conditions for Mycelial Biomass Production with Enhanced Antibacterial Activity of the Medicinal Macro Fungus *Xylaria* sp. Strain R006 against Drug Resistant Bacterial Pathogens." *Current Research in Environmental & Applied Mycology* 4, no. 1 (2014): 88–98.
22. Monika Moeder, Tomáš Cajthaml, Gábor Koeller, Pavla Erbanová, and Václav Šašek, "Structure Selectivity in Degradation and Translocation of Polychlorinated Biphenyls (Delor 103) with a *Pleurotus ostreatus* (Oyster Mushroom) Culture," *Chemosphere* 61, no. 9 (2005): 1370–1378.
23. P. Kalač and I. Stašková, "Concentrations of Lead, Cadmium, Mercury and Copper in Mushrooms in the Vicinity of a Lead Smelter," *Science of the Total Environment* 105 (1991): 109–119.
24. Alice Outwater, *Water: A Natural History*. (New York: Basic Books, 2008).
25. Michael M. Pollock, Morgan Heim, and Danielle Werner, "Hydrologic and Geomorphic Effects of Beaver Dams and Their Influence on Fishes," in *American Fisheries Society Symposium*, vol. 37, (2003), 213–233.
26. Ibid.
27. Robert J. Naiman, Carol A. Johnston, and James C. Kelley, "Alteration of North American Streams by Beaver," *BioScience* (1988): 753–762.
28. David F. Spencer, Liz Colby, and Gregory R. Norris, "An Evaluation of Flooding Risks Associated with Giant Reed (*Arundo donax*)," *Journal of Freshwater Ecology* 28, no. 3 (2013): 397–409.

Chapter 6: Everyone Gardens

1. John L. Sorenson and Carl L. Johannessen, *World Trade and Biological Exchanges Before 1492* (Bloomington, IN: iUniverse, 2009), 41.
2. Paul Tolstoy, "Cultural Parallels Between Southeast Asia and Mesoamerica in the Manufacture of Bark-cloth," *Transactions of the New York Academy of Sciences* 25 (1963): 646–662.
3. Robert G. Bednarik, "The Earliest Evidence of Ocean Navigation," *International Journal of Nautical Archaeology* 26, no. 3 (1997): 183–191.
4. Robert G. Bednarik, "Maritime Navigation in the Lower and Middle Palaeolithic," *Comptes Rendus de l'Académie des Sciences-Series IIA-Earth and Planetary Science* 328, no. 8 (1999): 559–563.
5. Bill Mollison, *Permaculture: A Designer's Manual* (Tasmania: Tagari Press, 1988), 97.

6. Caroline Roullier, Laure Benoit, Doyle B. McKey, and Vincent Lebot, "Historical Collections Reveal Patterns of Diffusion of Sweet Potato in Oceania Obscured by Modern Plant Movements and Recombination," *Proceedings of the National Academy of Sciences* 110, no. 6 (2013): 2205–2210.
7. Sorenson and Johannessen, *World Trade and Biological Exchanges Before 1492*, 77.
8. Ibid., 80.
9. Elsa Zardini, "*Madia sativa* Mol.(Asteraceae-Heliantheae-Madiinae): An Ethnobotanical and Geographical Disjunct," *Economic Botany* 46, no. 1 (1992): 34–44.
10. Missouri Native Seed Association, "Definition of Terms," accessed March 13, 2014, http://www.monativeseed.org/seedterminology.html.
11. Nancy J. Turner and Sandra Peacock, "Ethnobotanical Evidence for Plant Resource Management," in *Keeping It Living: Traditions of Plant Use and Cultivation on the Northwest Coast of North America*, ed. Douglas Deur and Nancy J. Turner (Seattle: University of Washington Press, 2005), 147.
12. Eduardo Goés Neves, James B. Petersen, Robert N. Bartone, and Michael J. Heckenberger, "The Timing of Terra Preta Formation in the Central Amazon: Archaeological Data from Three Sites," In *Amazonian Dark Earths: Explorations in Space and Time* (Springer Berlin Heidelberg, 2004), 125–134.
13. William M. Denevan, "The Pristine Myth: The Landscape of the Americas in 1492," *Annals of the Association of American Geographers* 82, no. 3 (1992): 369–385.
14. Darrell Addison Posey, "Indigenous Management of Tropical Forest Ecosystems: The Case of the Kayapo Indians of the Brazilian Amazon," *Agroforestry Systems* 3, no. 2 (1985): 139–158.
15. William M. Denevan, John M. Treacy, Janis B. Alcorn, Christine Padoch, Julie Denslow, and S. Flores Paitan, "Indigenous Agroforestry in the Peruvian Amazon: Bora Indian Management of Swidden Fallows," *Interciencia* 9, no. 6 (1984): 346–357.
16. Narciso Barrera-Bassols and Victor M. Toledo, "Ethnoecology of the Yucatec Maya: Symbolism, Knowledge and Management of Natural Resources," *Journal of Latin American Geography* 4, no. 1 (2005): 9–41.
17. Linda Christanty, Oekan S. Abdoellah, Gerald G. Marten, and Johan Iskandar, "Traditional Agroforestry in West Java: The Pekarangan (Homegarden) and Kebun-talun (Annual-Perennial Rotation) Cropping Systems," in *Traditional Agriculture in Southeast Asia: A Human Ecology Perspective*, ed. G. G. Marten (Boulder: Westview Press, 1986), 132–158.
18. Sidney Shetler, "Three Faces of Eden," in *Seeds of Change: A Quincentennial Commemoration*, ed. H. I. Viola and C. Margolis (Washington, DC: Smithsonian Institution, 1991), 231.
19. John Muir, *The Writings of John Muir*, vol. 1 (Boston: Houghton Mifflin, 1917), 132.
20. Robin Wall Kimmerer and Frank Kanawha Lake, "The Role of Indigenous Burning in Land Management," *Journal of Forestry* 99, no. 11 (2001): 36–41.
21. Muir, *Writings of John Muir*, 141.
22. Ibid.
23. Clark Spencer Larsen, "In the Wake of Columbus: Native Population Biology in the Postcontact Americas," *American Journal of Physical Anthropology* 37, no. S19 (1994): 109–154.
24. William M. Denevan, "The Pristine Myth: The Landscape of the Americas in 1492," *Annals of the Association of American Geographers* 82, no. 3 (1992): 369–385.
25. Francis Augustus MacNutt, *Bartholomew de las Casas: His Life, His Apostolate, and His Writings* (New York: Putnam, 1909), 314.
26. Timothy R. Pauketat, *Cahokia: Ancient America's Great City on the Mississippi* (New York: Penguin, 2009), 32.
27. Charles C. Mann, *1491: New Revelations of the Americas before Columbus* (New York: Random House LLC, 2005), 360.
28. Bob Zybach, PhD, NW Maps Company, e-mail message to author, August 15, 2010.

29. Gregory J. Nowacki and Marc D. Abrams, "The Demise of Dire and 'Mesophication' of Forests in the Eastern United States," *BioScience* 58, no. 2 (2008): 123–138.
30. Mann, *1491: New Revelations*, 360.
31. Erhard Rostlund, "The Geographic Range of the Historic Bison in the Southeast," *Annals of the Association of American Geographers* 50, no. 4 (1960): 395–407.
32. Alan K. Knapp, John M. Blair, John M. Briggs, Scott L. Collins, David C. Hartnett, Loretta C. Johnson, and E. Gene Towne, "The Keystone Role of Bison in North American Tallgrass Prairie," *BioScience* 49, no. 1 (1999): 39–50.
33. Kat Anderson, *Tending the Wild: Native American Knowledge and the Management of California's Natural Resources* (Berkeley: University of California Press, 2005), 4.
34. Fikret Berkes, *Sacred Ecology: Traditional Ecological Knowledge and Resource Management* (Oxford: Taylor & Francis 1999), 192.
35. Theodore A. Kennedy, Shahid Naeem, Katherine M. Howe, Johannes M. H. Knops, David Tilman, and Peter Reich, "Biodiversity as a Barrier to Ecological Invasion," *Nature* 417, no. 6889 (2002): 636–638.
36. Marla S. Hastings and Joseph M. DiTomaso, "Fire Controls Yellow Star Thistle in California Grasslands Test Plots at Sugarloaf Ridge State Park," *Ecological Restoration* 14, no. 2 (1996): 124–128.
37. Michele Merola-Zwartjes, "Biodiversity, Functional Processes, and the Ecological Consequences of Fragmentation in Southwestern Grasslands," in *Assessment of Grassland Ecosystem Conditions in the Southwestern United States,* vol. 1, Gen. Tech. Rep. RMRS-GTR-135-vol. 1, Ed. Deborah M. Finch (Fort Collins, CO: U.S. Department of Agriculture, Forest Service, Rocky Mountain Research Station, 2004): 49–85.
38. Joseph M. DiTomaso, Guy B. Kyser, and Marla S. Hastings, "Prescribed Burning for Control of Yellow Starthistle (*Centaurea solstitialis*) and Enhanced Native Plant Diversity," *Weed Science*, 47 (1999): 233–242.
39. David L. Strayer, Valerie T. Eviner, Jonathan M. Jeschke, and Michael L. Pace, "Understanding the Long-Term Effects of Species Invasions," *Trends in Ecology & Evolution* 21, no. 11 (2006): 645–651.
40. Dennis Martinez, "Introduction to Holistic Restoration Forestry" in *Forested Landscapes of Southwestern Oregon and Northern California*, Society for Ecological Restoration, January 2002, 8.
41. Victor King Chesnut, *Plants Used by the Indians of Mendocino County, California* (Washington, DC: US Government Printing Office, 1902).
42. P. J. Dailey, R. C. Graves, and J. L. Herring, "Survey of Hemiptera Collected on Common Milkweed, *Asclepias syriaca*, at One Site in Ohio," *Entomological News* 89, no. 7/8 (1978): 157–162.
43. Michael D. Reisner, James B. Grace, David A. Pyke, and Paul S. Doescher, "Conditions Favouring *Bromus tectorum* Dominance of Endangered Sagebrush Steppe Ecosystems," *Journal of Applied Ecology* 50, no. 4 (2013): 1039–1049.
44. K. T. Weber and B. S. Gokhale, "Effect of Grazing on Soil-Water Content in Semiarid Rangelands of Southeast Idaho," *Journal of Arid Environments* 75, no. 5 (2011): 464–470.
45. Luther Burbank, *How Plants Are Trained to Work for Man: Grafting and Budding,* vol. 2 (Boston: PF Collier & Son Company, 1921), 266–270.
46. Meriwether Lewis, William Clark, and Gary E. Moulton, eds. *The Definitive Journals of Lewis and Clark*, vol. 11 (Lincoln: University of Nebraska Press, 2003), 22.
47. Thomas C. Blackburn and Kat Anderson, "Introduction: Managing the Domesticated Environment," in *Before the Wilderness: Environmental Management by Native Californians*, ed. Thomas C. Blackburn and Kat Anderson (Menlo Park: Ballena Press, 1993), 19.
48. Nancy J. Turner and Sandra Peacock, "Ethnobotanical Evidence for Plant Resource Management," in *Keeping It Living: Traditions of Plant Use and Cultivation on the Northwest Coast of North America*, ed. Douglas Deur and Nancy J. Turner (Seattle: University of Washington Press, 2005), 145.

Chapter 7: Restoring Restoration

1. Joanna Macy as quoted by Mary Nurrie-Stearns in "Transforming Despair: An Interview with Joanna Macy," accessed June 8, 2014, http://www.personaltransformation.com/joanna_macy.html.
2. John W. Everest, James H. Miller, Donald M. Ball, and Mike Patterson, "Kudzu in Alabama History, Uses, and Control," Alabama A&M and Auburn Universities, Alabama Cooperative Extension System ANR-65, August 1999.
3. Betty Smith, *A Tree Grows in Brooklyn* (New York: Random House, 1992).
4. Marc C. Hoshovsky, "Element Stewardship Abstract for *Ailanthus altissima*," The Nature Conservancy, 1988.
5. Adewole L. Okunade, Rachel E. Bikoff, Steven J. Casper, Anna Oksman, Daniel E. Goldberg, and Walter H. Lewis, "Antiplasmodial Activity of Extracts and Quassinoids Isolated from Seedlings of *Ailanthus altissima* (Simaroubaceae)," *Phytotherapy Research* 17, no. 6 (2003): 675–677.
6. James Morris, "Ailanticulture – Its History and Commercial Relations," in *The Technologist: A Monthly Record of Science Applied to Art, Manufacture, and Culture*, vol II, ed. Peter Lund Simmons (Cambridge Scholars, 1862), 403.
7. Amol R. Dongre and Pradeep R. Deshmukh, "Farmers' Suicides in the Vidarbha Region of Maharashtra, India: A Qualitative Exploration of their Causes," *Journal of Injury and Violence Research* 4, no. 1 (2012): 2.
8. Miguel A. Altieri, "Agroecology, Small Farms, and Food Sovereignty," *Monthly Review* 61, no. 3 (2009): 102–113.
9. John Russell Smith, *Tree Crops: A Permanent Agriculture* (Washington, DC: Island Press, 1987), 351.
10. T. E. Crews, "Perennial Crops and Endogenous Nutrient Supplies," *Renewable Agriculture and Food Systems* 20, no. 01 (2005): 25–37.
11. Food and Agriculture Organization (FAO), "Perennial Agriculture: Landscape Resilience for the Future," http://www.fao.org/fileadmin/templates/agphome/documents/scpi/PerennialPolicyBrief.pdf.
12. Jonathan A. Patz, Mike Hulme, Cynthia Rosenzweig, Timothy D. Mitchell, Richard A. Goldberg, Andrew K. Githeko, Subhash Lele, Anthony J. McMichael, and David Le Sueur, "Climate Change (Communication Arising): Regional Warming and Malaria Resurgence," *Nature* 420, no. 6916 (2002): 627–628.
13. Lisamarie Windham and Laura A. Meyerson, "Effects of Common Reed (*Phragmites australis*) Expansions on Nitrogen Dynamics of Tidal Marshes of the Northeastern US," *Estuaries* 26, no. 2 (2003): 452–464.

Chapter 8: Putting Permaculture to Work in Restoration

1. Bill Mollison, *Permaculture: A Designer's Manual* (Tyalgum, Australia: Tagari Publications, 1988), ix.
2. Shunryu Suzuki, *Zen Mind, Beginner's Mind* (Boston: Shambhala Publications, 2010), 1.
3. Lauren Samantha Urgenson, "Ecological Consequences of Knotweed Invasion in Riparian Forests," master's thesis, University of Washington, 2006, 5.
4. Samuel J. McNaughton, "Diversity and Stability of Ecological Communities: A Comment on the Role of Empiricism in Ecology," *American Naturalist* (1977): 515–525.
5. William F. Laurance, "Edge Effects in Tropical Forest Fragments: Application of a Model for the Design of Nature Reserves," *Biological Conservation* 57, no. 2 (1991): 205–219.
6. Willem C. Beets, *Multiple Cropping and Tropical Farming Systems* (Farnham, UK: Gower Publishing Co., 1982).
7. Bill Mollison and Reny Mia Slay, *Introduction to Permaculture* (Tyalgum, Australia: Tagari Publications, 1991), 177.

Index

C

Index

H

R

Y

Z

About the Author

Tao Orion is a permaculture designer, teacher, homesteader, and mother living in the southern Willamette Valley of Oregon. She teaches permaculture design at Oregon State University and at Aprovecho, a 40-acre nonprofit sustainable-living educational organization. Tao consults on holistic farm, forest, and restoration planning through Resilience Permaculture Design, LLC. She holds a degree in agroecology and sustainable agriculture from UC Santa Cruz, and her interest in restoration was piqued when studying botany, wildcrafting, and herbalism at the Columbines School of Botanical Studies in Eugene, Oregon. She has a keen interest in integrating the disciplines of organic agriculture, sustainable land-use planning, ethnobotany, and ecosystem restoration in order to create beneficial social, economic, and ecological outcomes. When she is not writing, she is busy keeping up with her toddler and wrangling a diverse array of plants and animals on her 6.5-acre homestead, Viriditas Farm.

Chelsea Green Publishing is committed to preserving ancient forests and natural resources. We elected to print this title on paper containing 100% postconsumer recycled paper, processed chlorine-free. As a result, for this printing, we have saved:

46 Trees (40' tall and 6-8" diameter)
21,330 Gallons of Wastewater
21 million BTUs Total Energy
1,428 Pounds of Solid Waste
3,933 Pounds of Greenhouse Gases

Chelsea Green Publishing made this paper choice because we are a member of the Green Press Initiative, a nonprofit program dedicated to supporting authors, publishers, and suppliers in their efforts to reduce their use of fiber obtained from endangered forests. For more information, visit www.greenpressinitiative.org.

Environmental impact estimates were made using the Environmental Defense Paper Calculator. For more information visit: www.papercalculator.org.